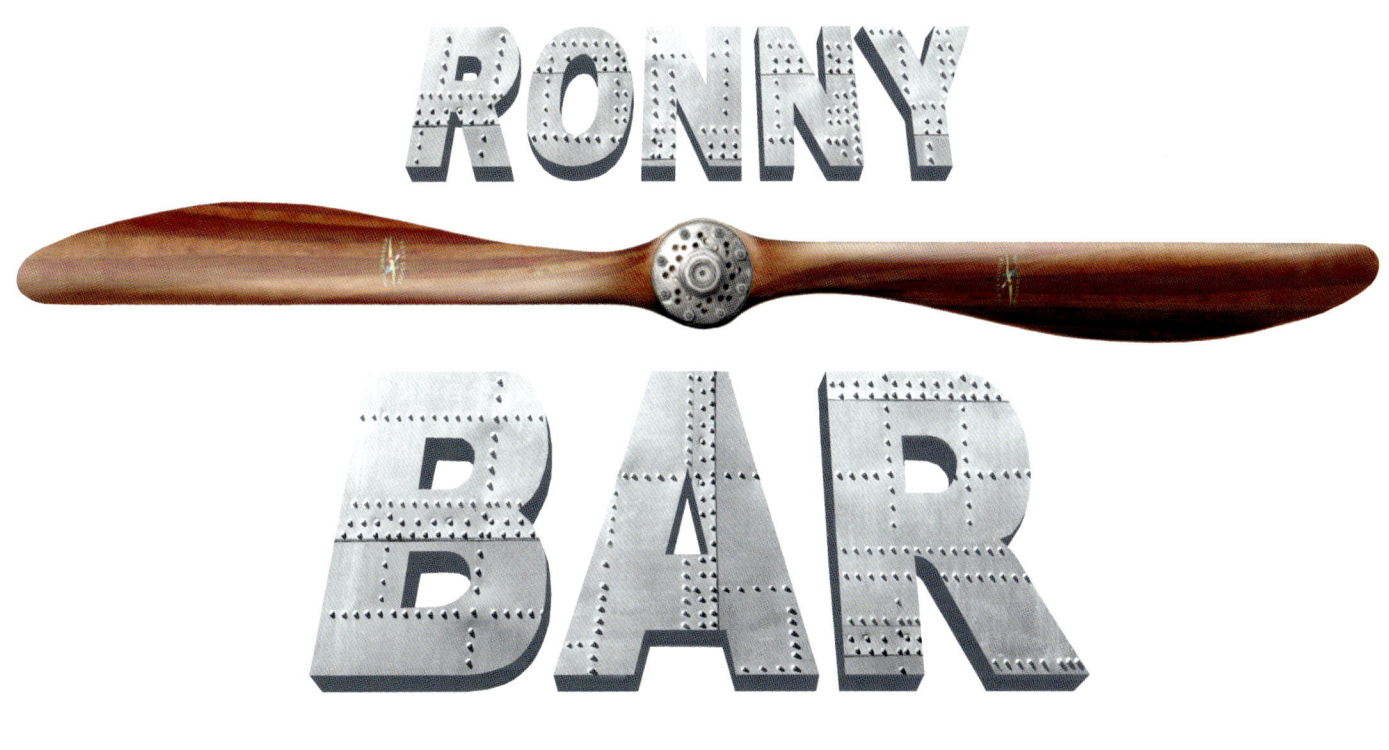

RONNY BAR
PROFILES

Ronny Bar
Profiles
Spitfire
The Merlin Variants

CONTENTS

Introduction	7	Spitfire Mk VII	109
Foreword	9	Spitfire Mk VIII	117
Spitfire Prototype	11	Spitfire Mk IX	137
Spitfire Mk I	15	Spitfire Mk XVI	177
Spitfire Mk II	47	PR Spitfires	195
Spitfire Mk V	63	Seafire	215
Spitfire Mk VI	105		

INTRODUCTION

"The Spitfire is the most beautiful airplane of all time"

I am absolutely convinced that this statement of mine is shared by thousands of aviation enthusiasts around the world.

When I was a little boy, I first learned of the existence of the Spitfire through a comic strip: Lúpin (a Spanishization of 'Looping') Lupin was a young man who owned a biplane, some sort of an SE5a converted to a two-seater, his grandfather fought in the RFC in World War I and his father flew Spitfires for the RAF in World War II (like several thousand young Argentines who fought on the British side in both wars).

After that I began to learn more and more about this famous aircraft, and over the years I have come to make countless drawings and build dozens of models of the Spitfire. Until one day, on my first visit to the RAF Museum at Hendon, rounding the corner of one of its halls, I came face to face with a Spitfire Mk I in glorious Battle of Britain colours… I was shocked! It was as though I was a priest standing in front of the Holy Grail! I remember running to buy a phone card and calling my wife in Buenos Aires to tell her that I had met a Spitfire! Of course, her response was: "You met who…?" (laughter)

A couple of weeks later, going to an air show at Duxford, I remember getting off the bus at the airfield gates, hearing a peculiar sound, the unmistakable roar of a Merlin engine. When I looked up, there she was, those lovely elliptical wings finished in PRU Blue, doing aerobatics… She was such a beauty! The Marilyn Monroe of airplanes!

Many years later, already established as a professional aircraft profile artist, and having worked on dozens of types of aircraft, mostly from WWI, but also quite a few from WWII, and even some from the interwar period, I still hadn't had the chance of profiling a Spitfire… Then, when working for a German magazine specializing in WWII German planes, the editor was so happy with my work that one day he asked me which was my favourite plane, and which one I would like to illustrate.

Without hesitation I replied: "Spitfire!" (shades of Adolph Galland…!) He said: "But we make do German planes!" So I replied: "Well, a *beute* (captured) Spitfire…" But unfortunately it never happened.

Then, another day, after Peter Jackson's legendary Wingnut Wings Models closed shop, this company's ex-general manager, my good friend Richard Alexander, contacted me about Kotare Models. This is a New Zealand-based kits manufacturing company that he and other ex-Wingnuts guys created to produce WWII fighters models, and he asked me if I would be willing to join… Obviously I said a huge "YES!" and that then I would finally get a chance to work on the Spit, to which he answered: "Maybe… Probably you will." Well to my surprise, and delight, the first kit they released was an amazing 1/32 scale model of the Spitfire Mk I.

After finishing the required profiles for the kit, I found myself doing more and more profiles of different marks of Spitfires in my spare time between other jobs, and just for fun, for love of the airplane! Eventually I found I had so many Spitfire profiles that I couldn't help but think that they definitely deserved a book! I focused on the major variants of the Merlin-powered Spitfires that went into production. The Griffon-powered variants are a whole different story (if not even a different aircraft)!

I realised that whatever I could say about this airplane had already been written in countless books and articles by people much more authorised than me on this subject, and that what readers would really want when buying this book would be the profiles, so I decided to base this book only on them, with a minimum of data on each subject.

A few of these profiles have been published in Kotare Models, Amber Books and Wingleader before, but the vast majority were made especially for this book. So here it is… I hope you enjoy going through these pages as much as I enjoyed making these profiles.

Ronny Bar

FOREWORD

It is a great pleasure for me to be able to write the foreword for this new profile art book by my long-time good friend, Ronny Bar. I first met Ronny nearly twenty years ago during the very early days of the Wingnut Wings model kits company and his subsequent visits to New Zealand.

If you're holding this book, it is not unreasonable to assume that you and I share at least a couple of things in common. The first thing that we share would be an appreciation of the Supermarine Spitfire, which is almost certainly the most beautiful aircraft ever to go into mass production.

The Supermarine Spitfire prototype first flew in March 1936 and was a quantum leap forward compared with the biplane fighters equipping the RAF squadrons at that time. The first production aircraft was delivered in mid-May 1938 and over 22,000 had been built in two-dozen major variants by the time production ceased 10 years later. Numerous engine, airframe and armament improvements were made to the Spitfire over this time, and Ronny's book covers the main Rolls-Royce Merlin powered variants such as the 1,030hp Mk I, 1,150hp Mk II, 1,470hp Mk V and 1,720hp Mk IX. My particular favourite has always been the Battle of Britain era Mk I. Hopefully Ronny can be convinced to produce a follow up volume including the Rolls-Royce Griffon powered Spitfires at some time in the future.

The second thing that we have in common is an appreciation for the beautiful profile artwork of Ronny Bar. No matter what the subject, Ronny's profile art is always beautifully detailed, and he really puts his heart into trying to get every little aspect correct, which is a characteristic that I can of course only admire. That's not to say that we always agreed on what was 'correct', but the process of critically appraising all available reference sources in our efforts to try and present an illustration as close as possible to 'the truth' was always a very rewarding experience.

Ronny must have illustrated thousands of profiles for me at Wingnut Wings and I'm very glad to be able to continue our partnership now with Kotare Models (as well as with the most recent Wingleader Supermarine Spitfire Mk I/II Special Edition book) where Ronny continues his perfectionist goals.

It is remarkable how Ronny has been able to illustrate so many Merlin-powered Spitfires after being ensconced for so long in the world of WWI aircraft. There are so many in fact that I haven't even had time to properly digest them all, so I'm really looking forward to being able to peruse this book at leisure and properly admire Ronny's beautiful illustrations of the most beautiful aircraft of all time. They might even provide me with some ideas for the future.

Richard Alexander
November 2023

Spitfire Prototype

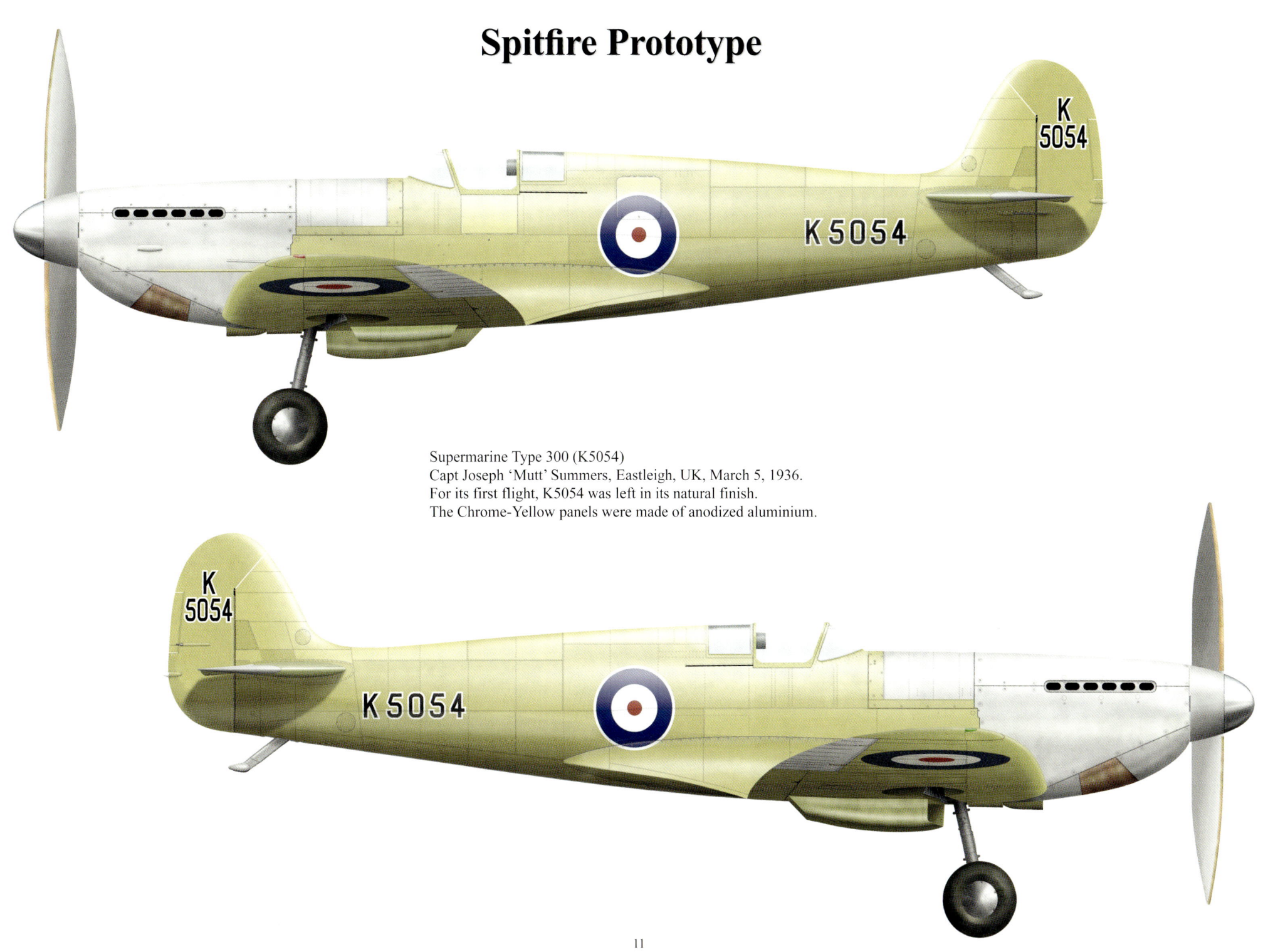

Supermarine Type 300 (K5054)
Capt Joseph 'Mutt' Summers, Eastleigh, UK, March 5, 1936.
For its first flight, K5054 was left in its natural finish.
The Chrome-Yellow panels were made of anodized aluminium.

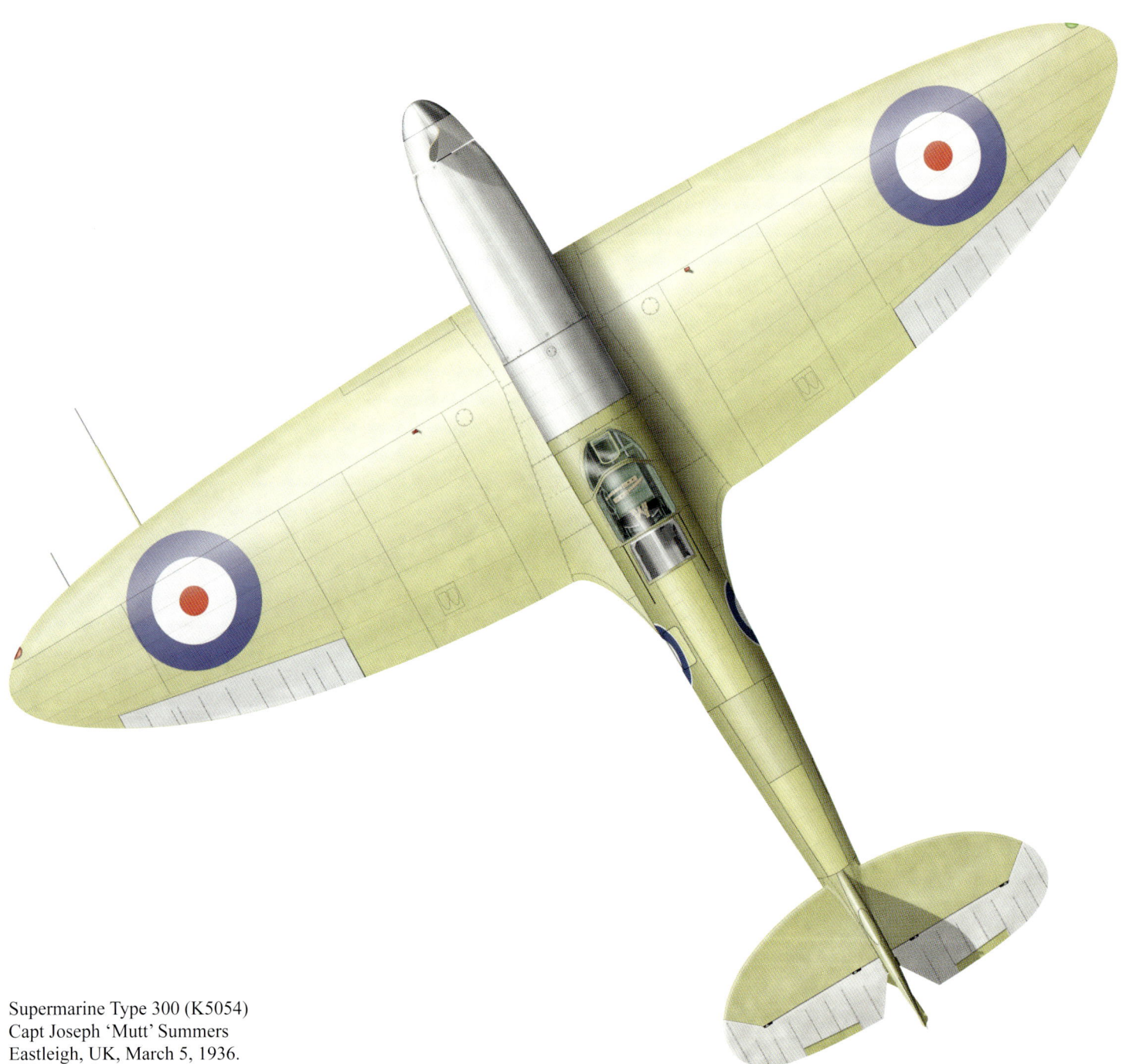

Supermarine Type 300 (K5054)
Capt Joseph 'Mutt' Summers
Eastleigh, UK, March 5, 1936.

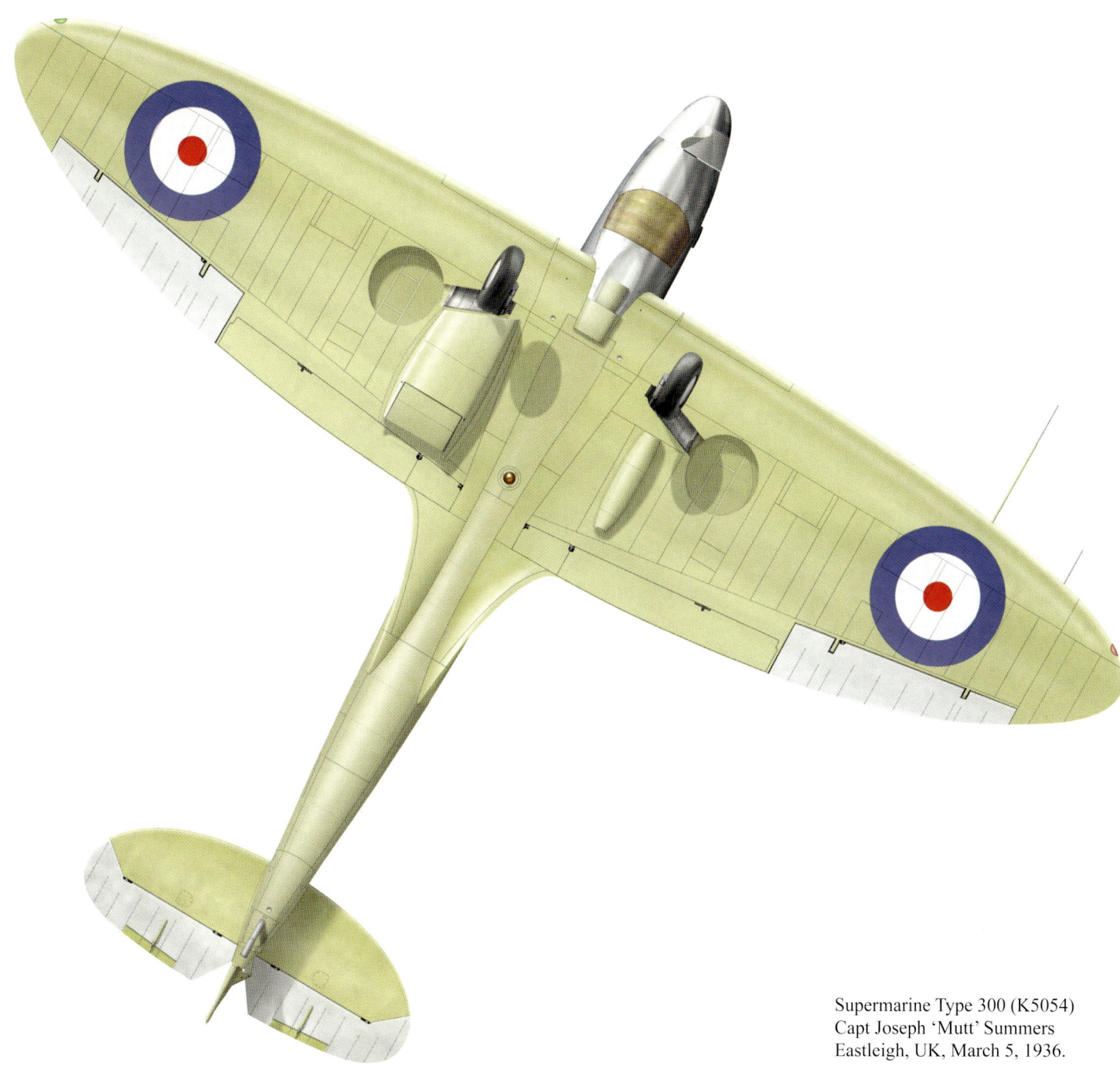

Supermarine Type 300 (K5054)
Capt Joseph 'Mutt' Summers
Eastleigh, UK, March 5, 1936.

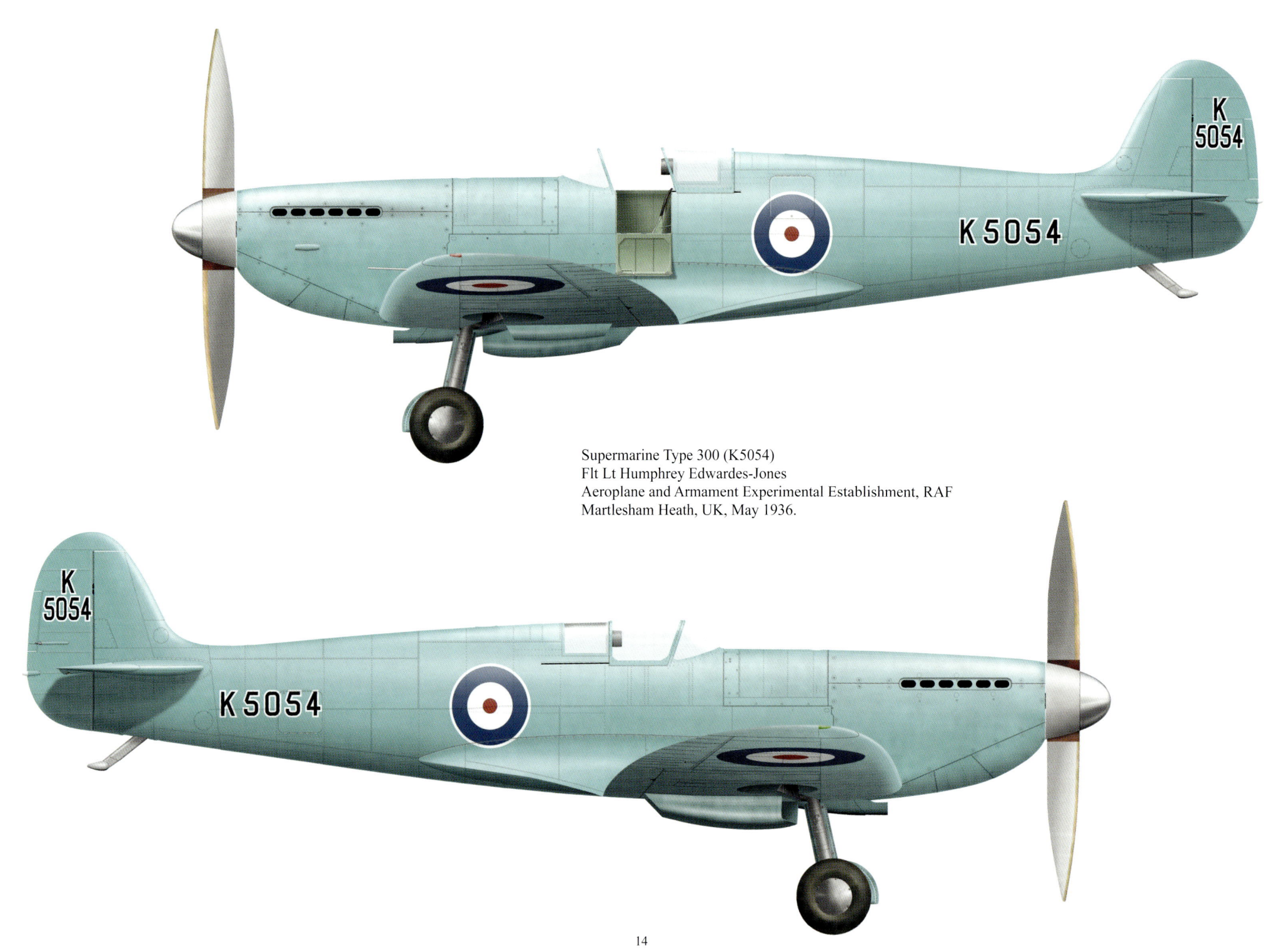

Supermarine Type 300 (K5054)
Flt Lt Humphrey Edwardes-Jones
Aeroplane and Armament Experimental Establishment, RAF
Martlesham Heath, UK, May 1936.

Spitfire Mark I

Supermarine Spitfire Mk I (K9795)
19 Squadron, RAF
Duxford, UK, October 1938.

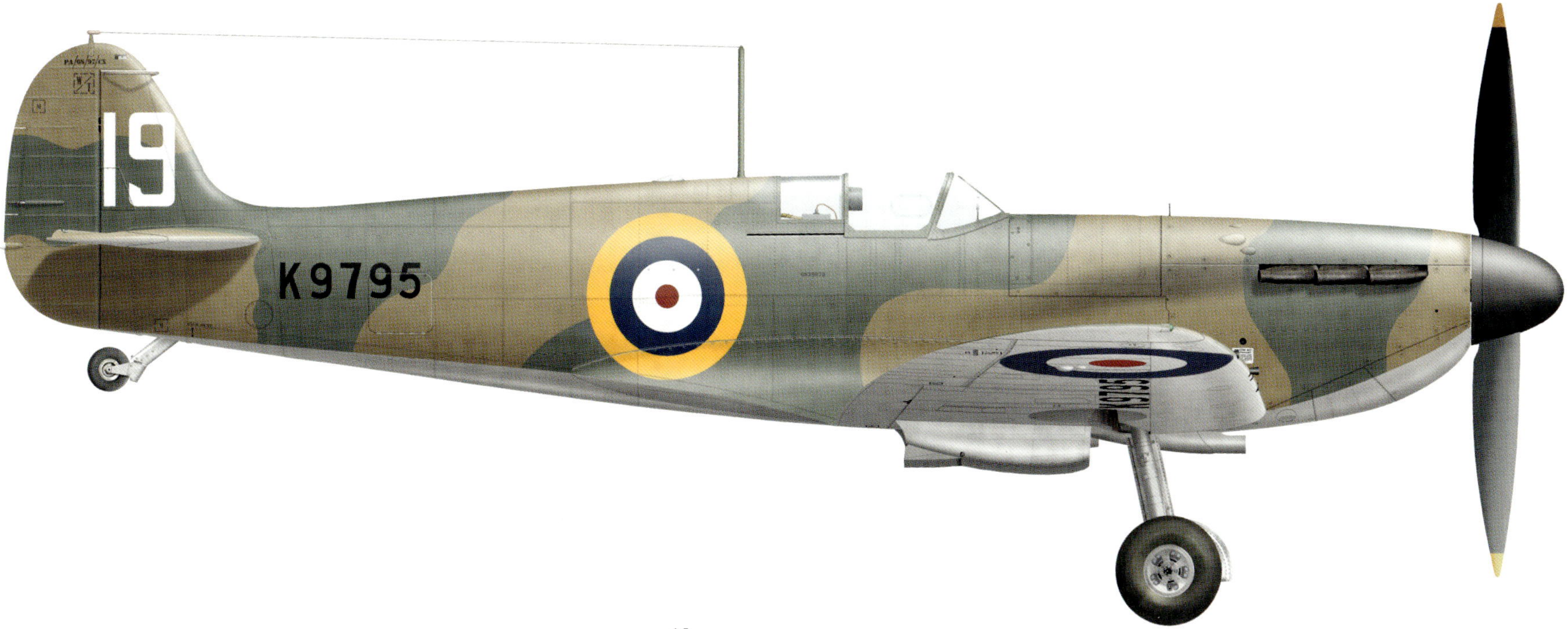

Supermarine Spitfire Mk I (K9795)
19 Squadron, RAF
Duxford, UK, October 1938.

Supermarine Spitfire Mk I (K9795)
19 Squadron, RAF
Duxford, UK, October 1938.

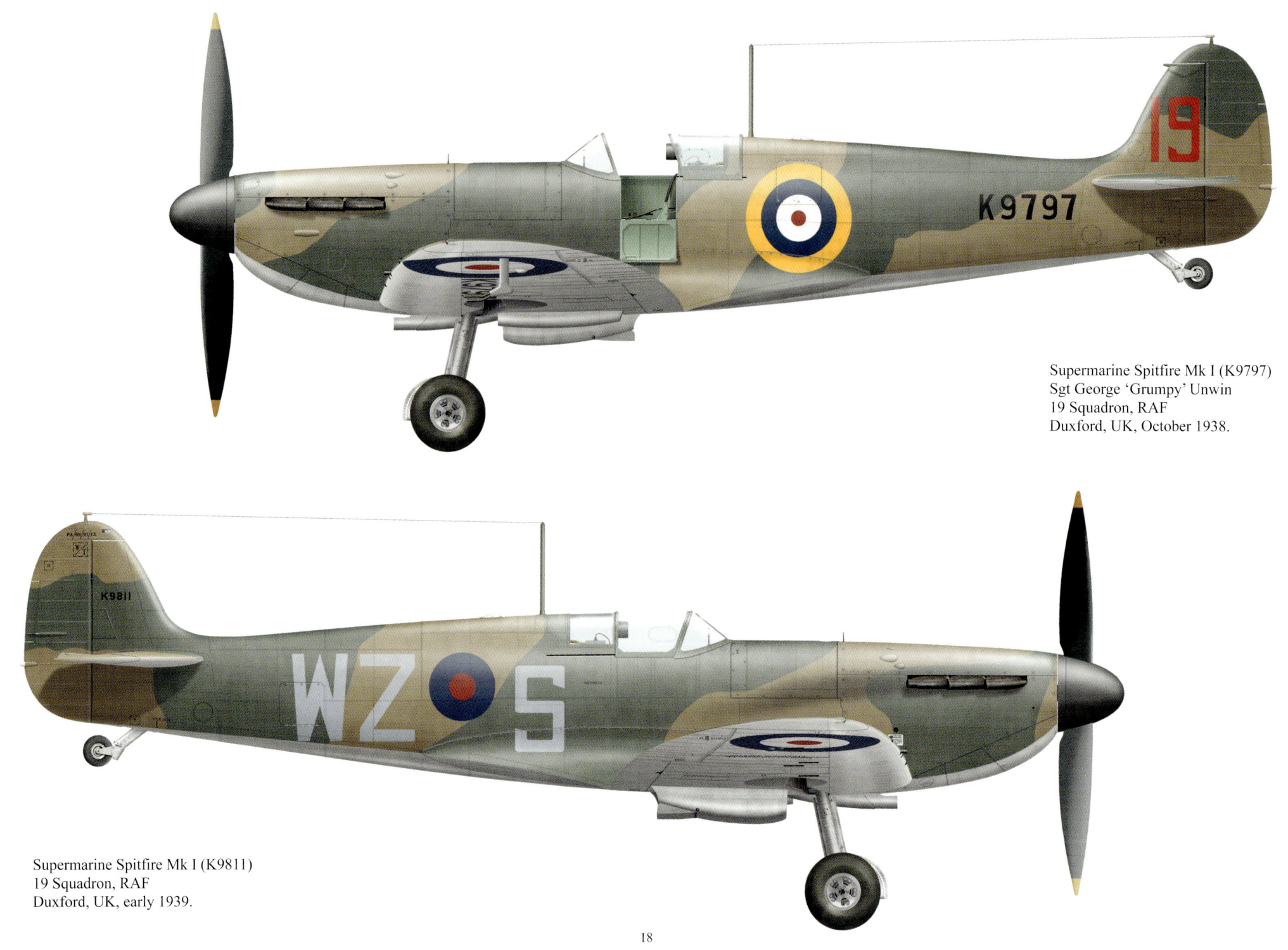

Supermarine Spitfire Mk I (K9797)
Sgt George 'Grumpy' Unwin
19 Squadron, RAF
Duxford, UK, October 1938.

Supermarine Spitfire Mk I (K9811)
19 Squadron, RAF
Duxford, UK, early 1939.

Supermarine Spitfire Mk I (K9798)
19 Squadron, RAF
Duxford, UK, April 1939.

Supermarine Spitfire Mk I (K9798)
19 Squadron, RAF
Duxford, UK, April 1939.

Supermarine Spitfire Mk I (K9798)
19 Squadron, RAF
Duxford, UK, April 1939.

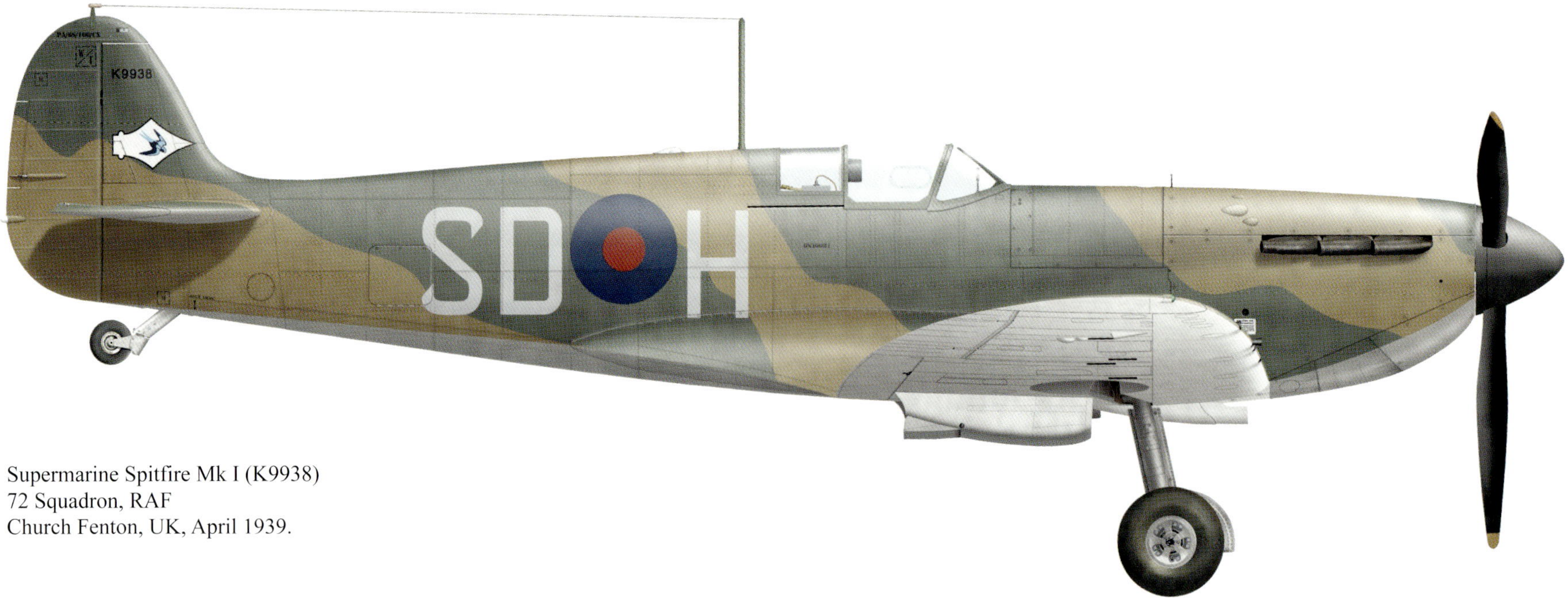

Supermarine Spitfire Mk I (K9938)
72 Squadron, RAF
Church Fenton, UK, April 1939.

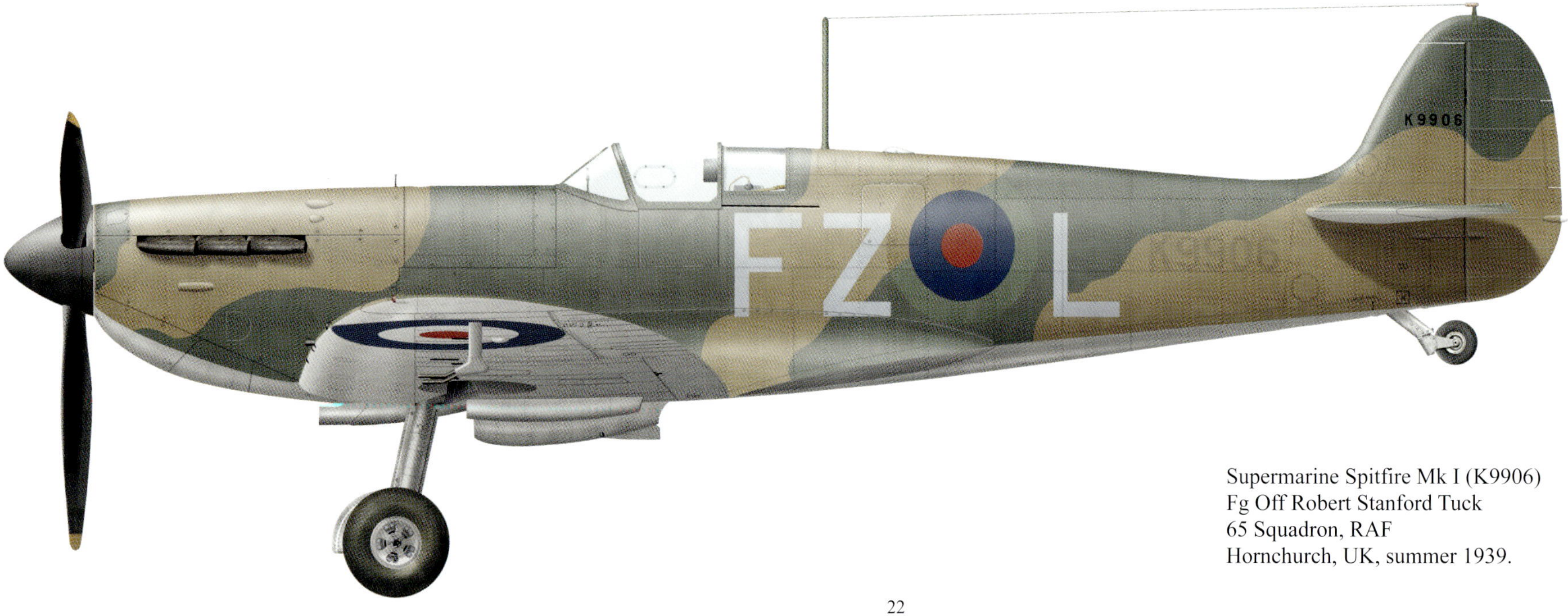

Supermarine Spitfire Mk I (K9906)
Fg Off Robert Stanford Tuck
65 Squadron, RAF
Hornchurch, UK, summer 1939.

Supermarine Spitfire Mk I (L1070)
Flt Lt Patrick Gifford
603 (City of Edinburgh) Squadron, RAF
Turnhouse, UK, October 1939.

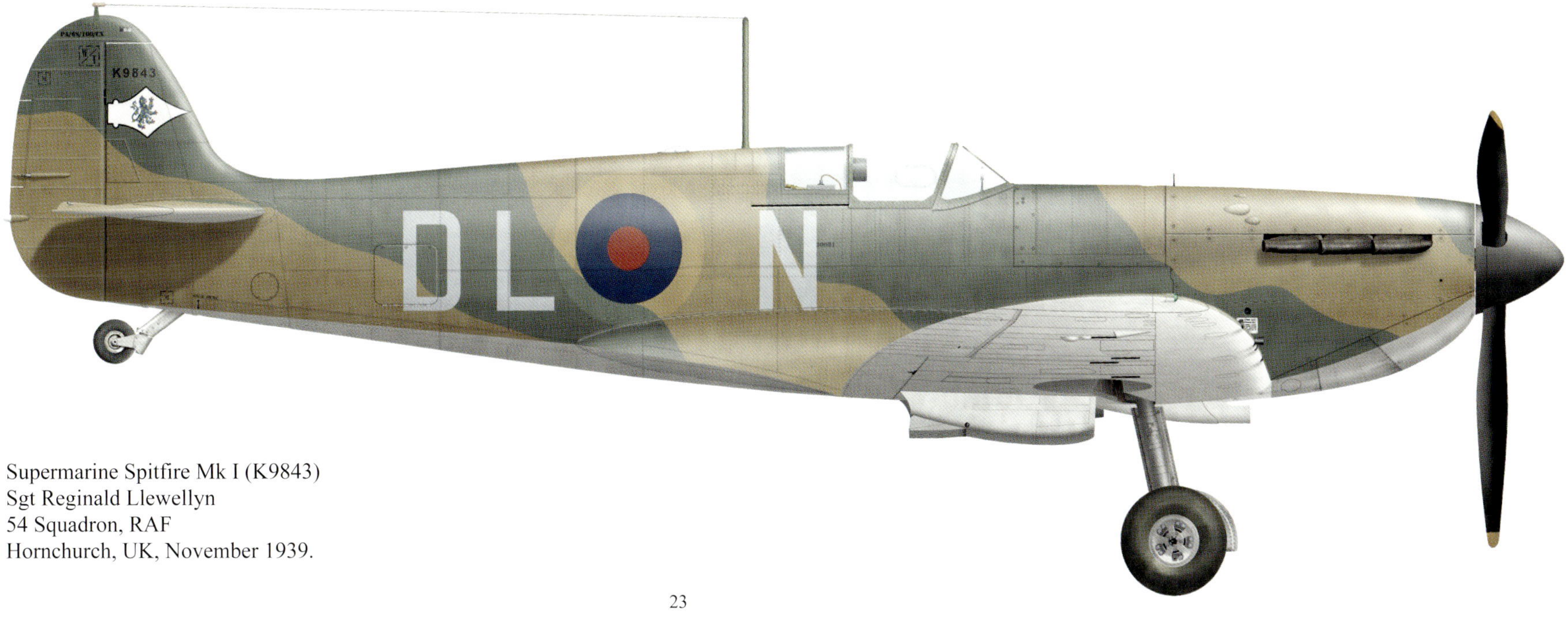

Supermarine Spitfire Mk I (K9843)
Sgt Reginald Llewellyn
54 Squadron, RAF
Hornchurch, UK, November 1939.

Supermarine Spitfire Mk I (K9955)
Fg Off Archibald McKellar
602 (City of Glasgow) Squadron, RAF
Drem, UK, March 1940.

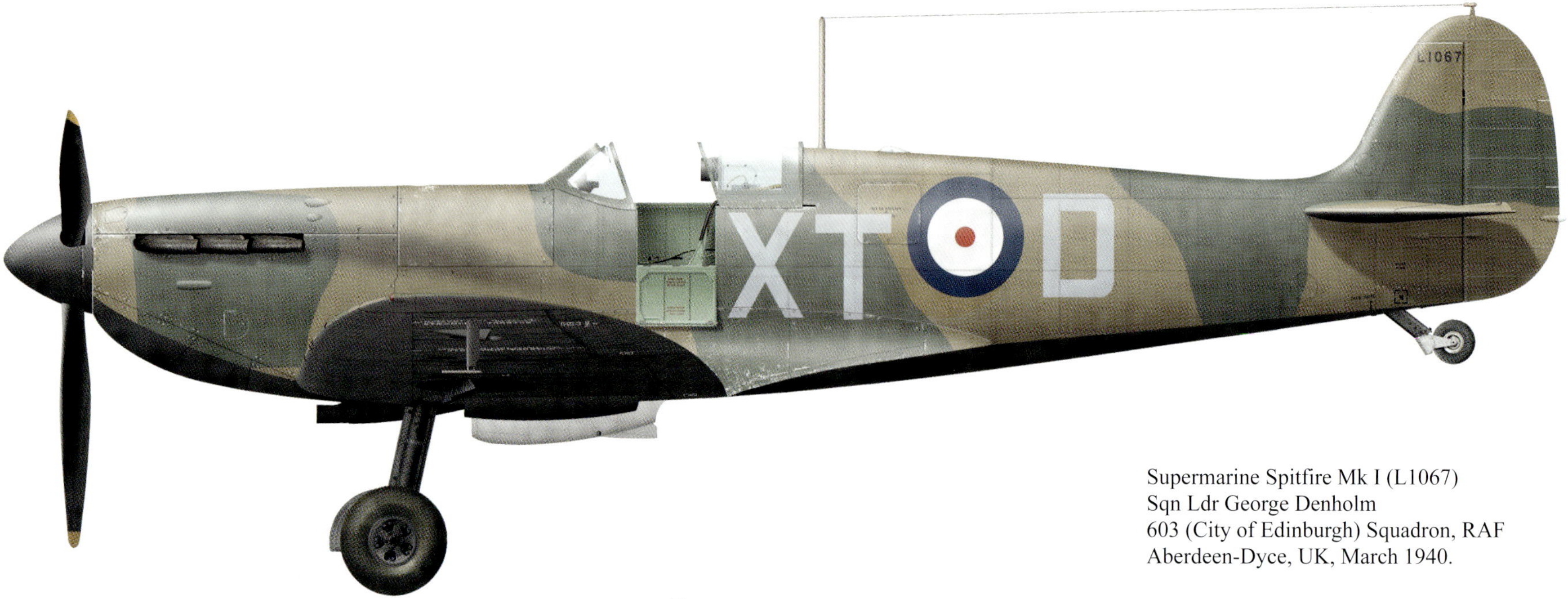

Supermarine Spitfire Mk I (L1067)
Sqn Ldr George Denholm
603 (City of Edinburgh) Squadron, RAF
Aberdeen-Dyce, UK, March 1940.

Supermarine Spitfire Mk I (N3180)
Plt Off Alan Deere
54 Squadron, RAF
Hornchurch, UK, May 1940.

Supermarine Spitfire Mk I (N3180)
Plt Off Alan Deere
54 Squadron, RAF
Hornchurch, UK, May 1940.

Supermarine Spitfire Mk I (N3180)
Plt Off Alan Deere
54 Squadron, RAF
Hornchurch, UK, May 1940.

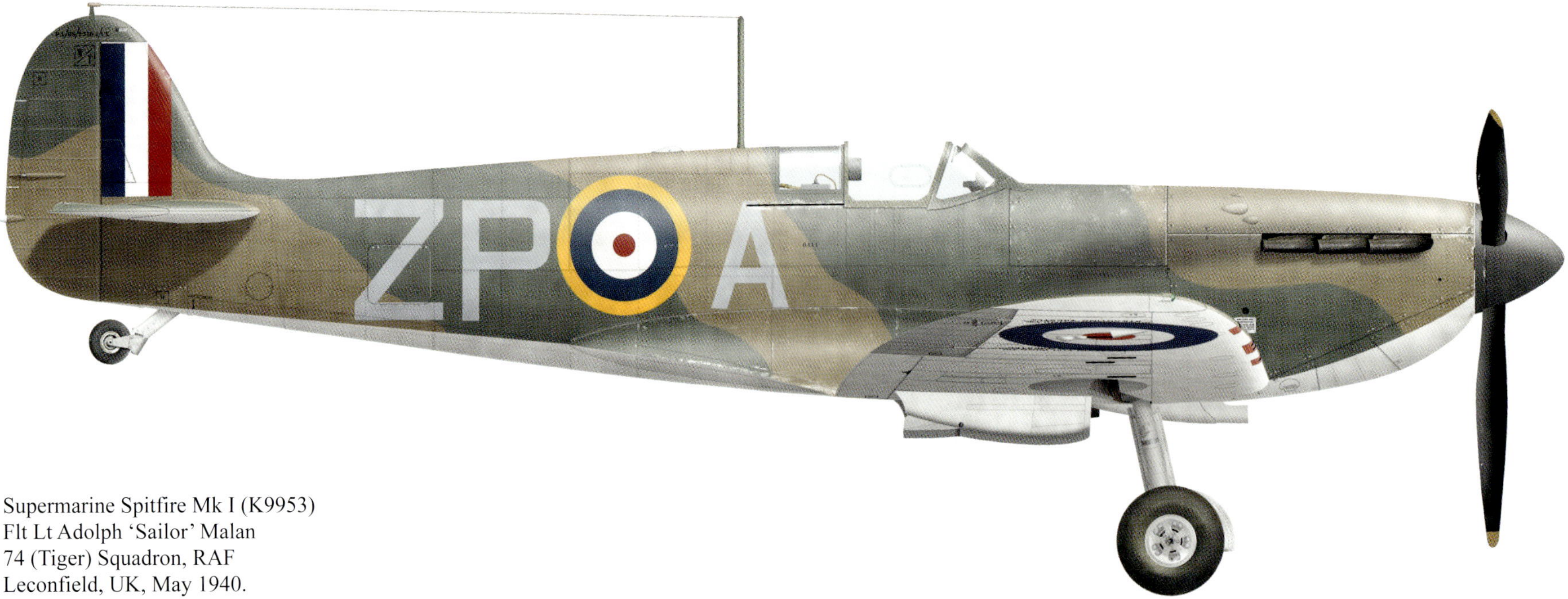

Supermarine Spitfire Mk I (K9953)
Flt Lt Adolph 'Sailor' Malan
74 (Tiger) Squadron, RAF
Leconfield, UK, May 1940.

Supermarine Spitfire Mk I (P9374)
Plt Off Desmond Williams
92 Squadron, RAF
Croydon, UK, May 1940.

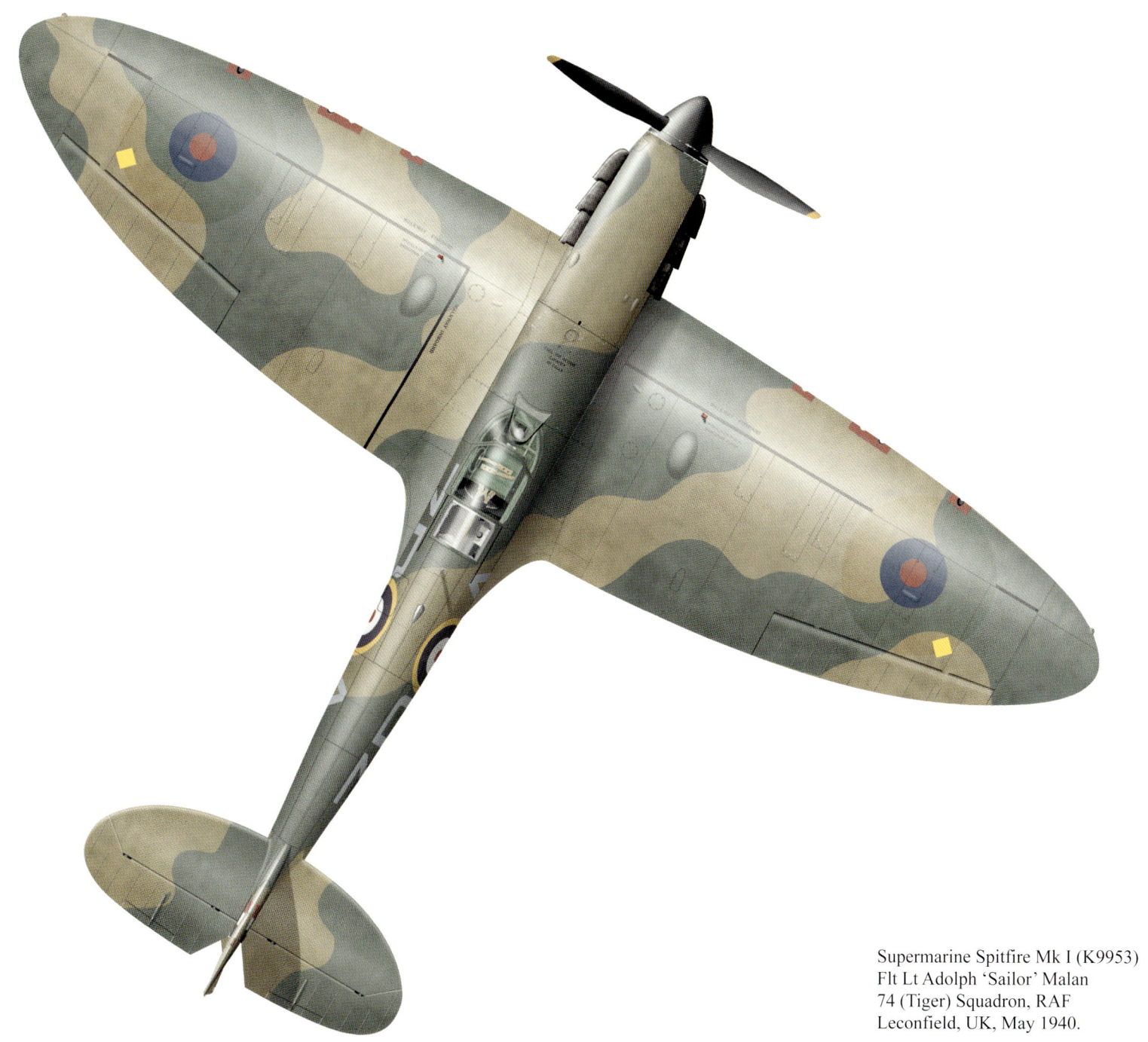

Supermarine Spitfire Mk I (K9953)
Flt Lt Adolph 'Sailor' Malan
74 (Tiger) Squadron, RAF
Leconfield, UK, May 1940.

Supermarine Spitfire Mk I (N3250)
Plt Off Allan Wright
92 Squadron, RAF
Croydon, UK, June 1940.

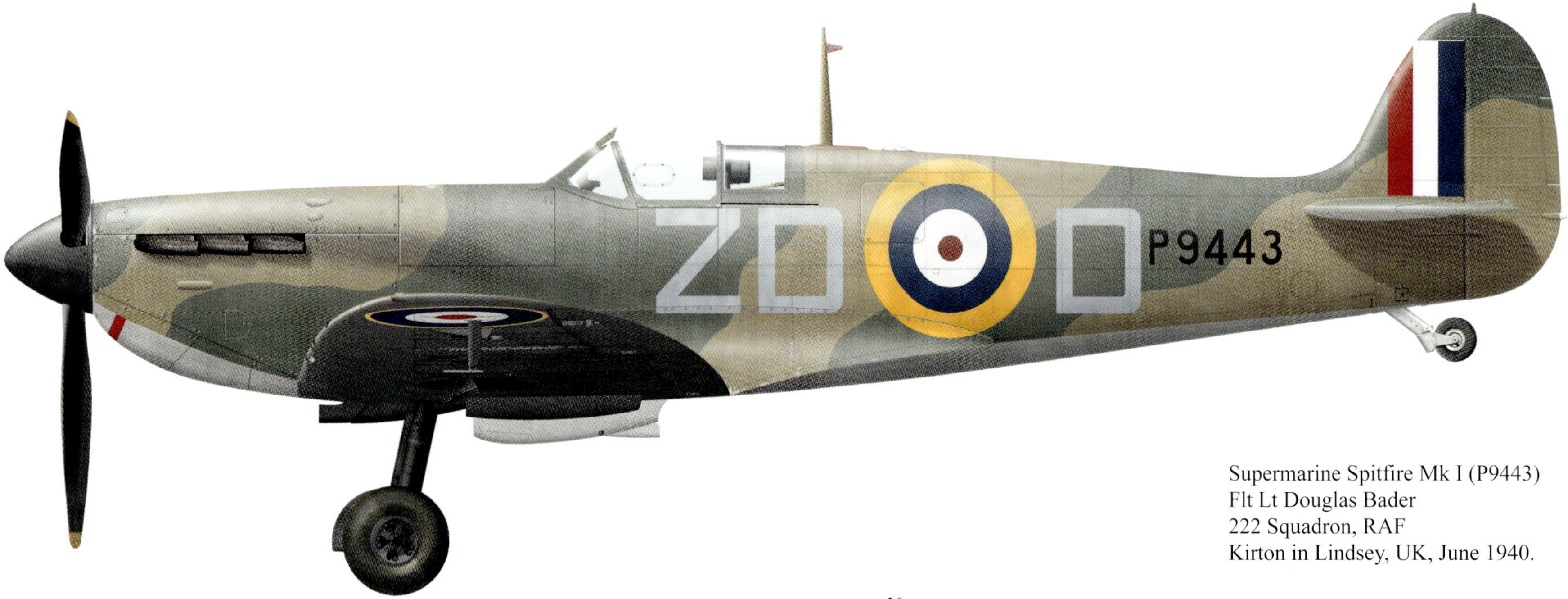

Supermarine Spitfire Mk I (P9443)
Flt Lt Douglas Bader
222 Squadron, RAF
Kirton in Lindsey, UK, June 1940.

Supermarine Spitfire Mk I (P9495)
Sgt Stanley J. Arnfield
610 (County of Chester) Squadron, RAF
Biggin Hill, UK, July 1940.

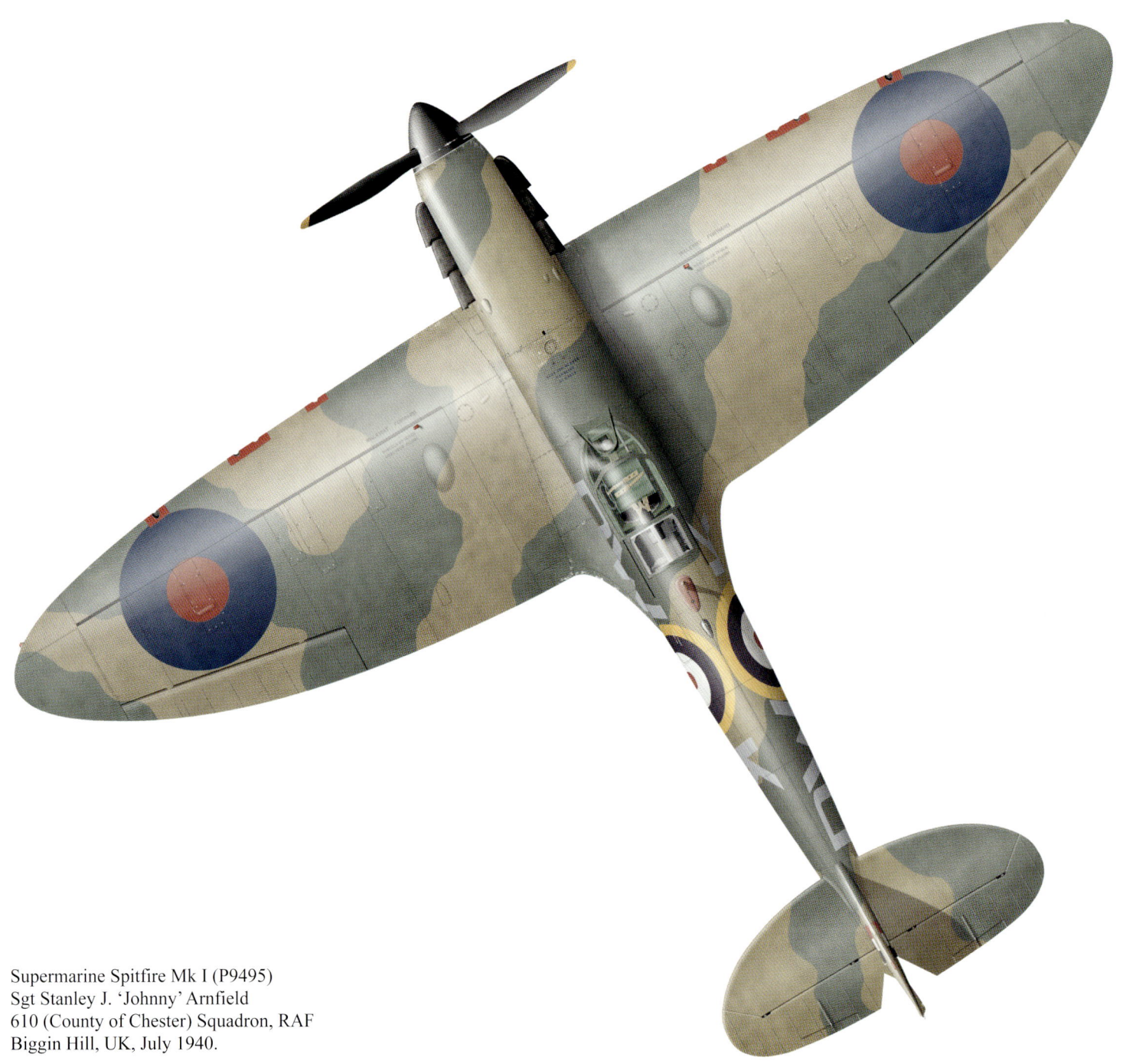

Supermarine Spitfire Mk I (P9495)
Sgt Stanley J. 'Johnny' Arnfield
610 (County of Chester) Squadron, RAF
Biggin Hill, UK, July 1940.

Supermarine Spitfire Mk I (P9495)
Sgt Stanley J. 'Johnny' Arnfield
610 (County of Chester) Squadron, RAF
Biggin Hill, UK, July 1940.

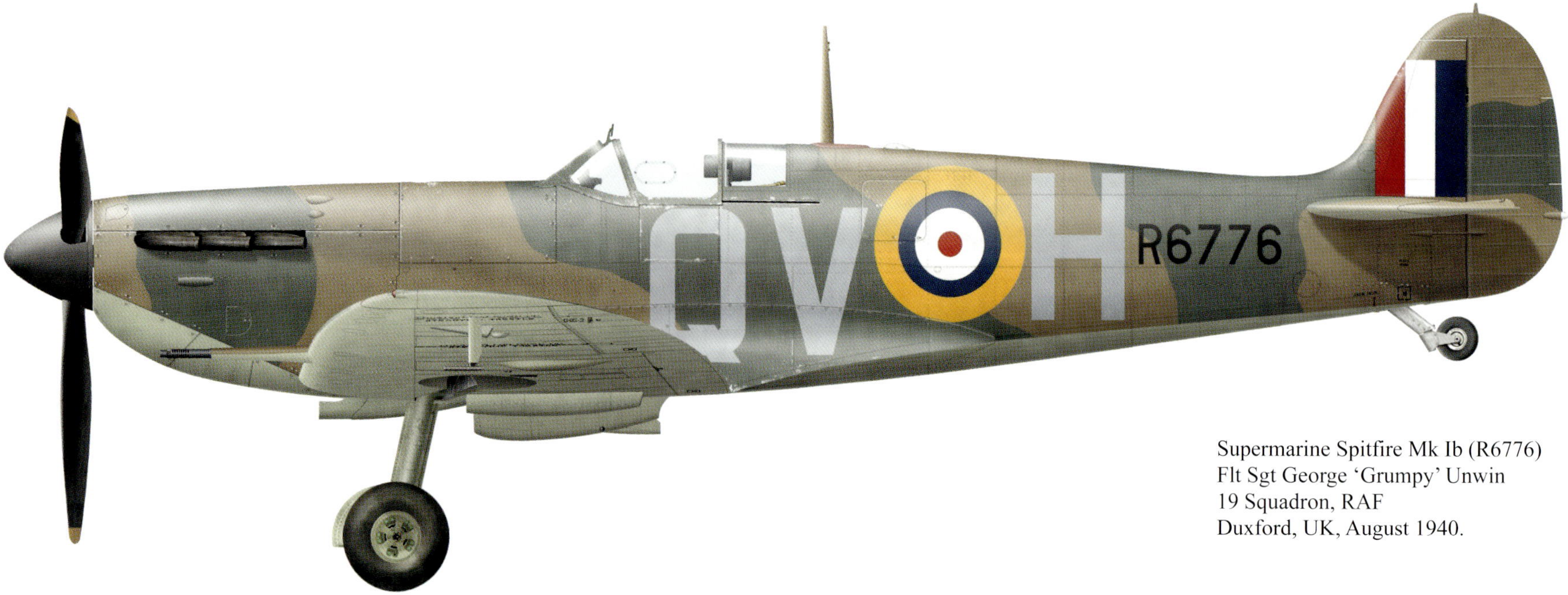

Supermarine Spitfire Mk Ib (R6776)
Flt Sgt George 'Grumpy' Unwin
19 Squadron, RAF
Duxford, UK, August 1940.

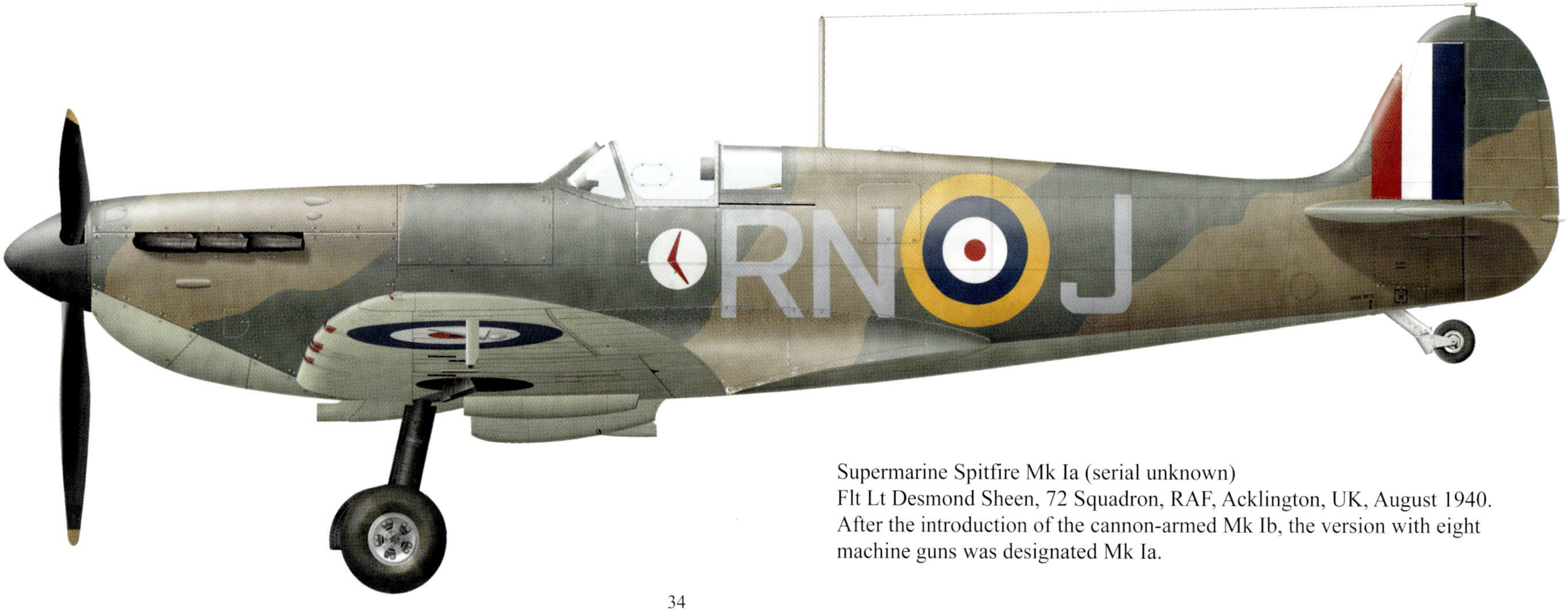

Supermarine Spitfire Mk Ia (serial unknown)
Flt Lt Desmond Sheen, 72 Squadron, RAF, Acklington, UK, August 1940.
After the introduction of the cannon-armed Mk Ib, the version with eight
machine guns was designated Mk Ia.

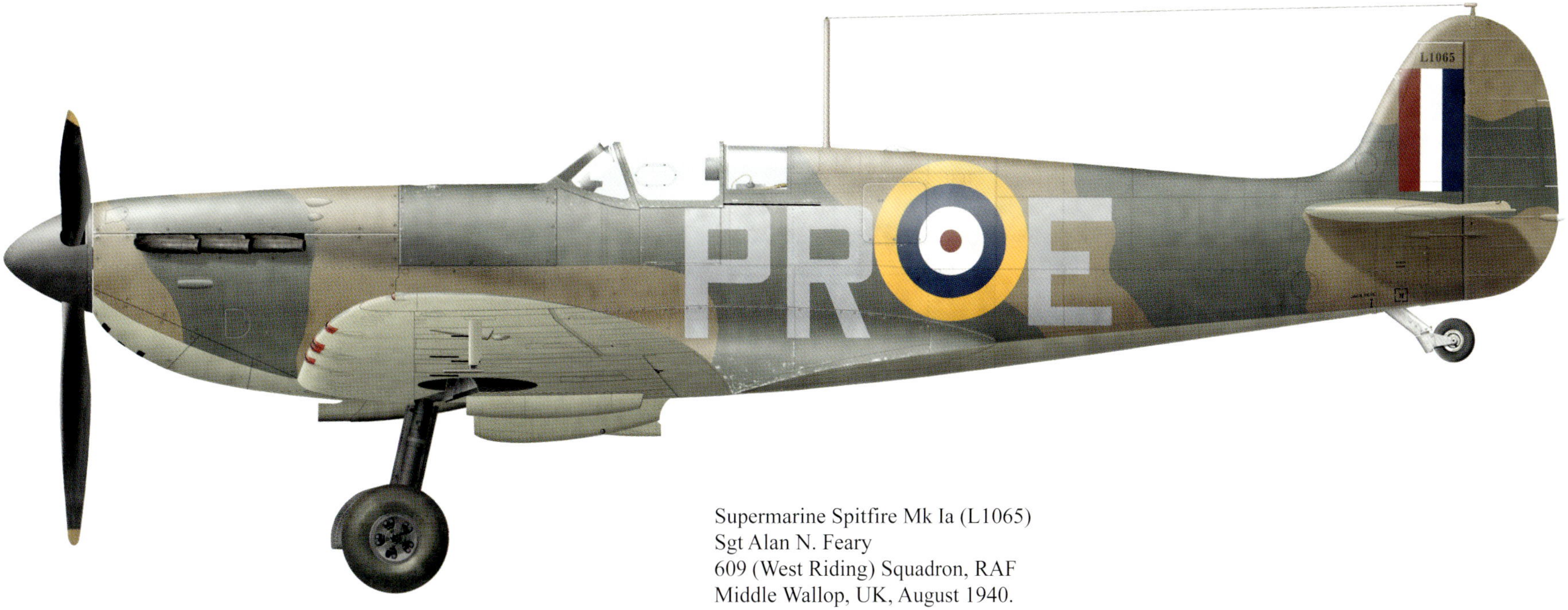

Supermarine Spitfire Mk Ia (L1065)
Sgt Alan N. Feary
609 (West Riding) Squadron, RAF
Middle Wallop, UK, August 1940.

Supermarine Spitfire Mk Ia (L1065)
Sgt Alan N. Feary
609 (West Riding) Squadron, RAF
Middle Wallop, UK, August 1940.

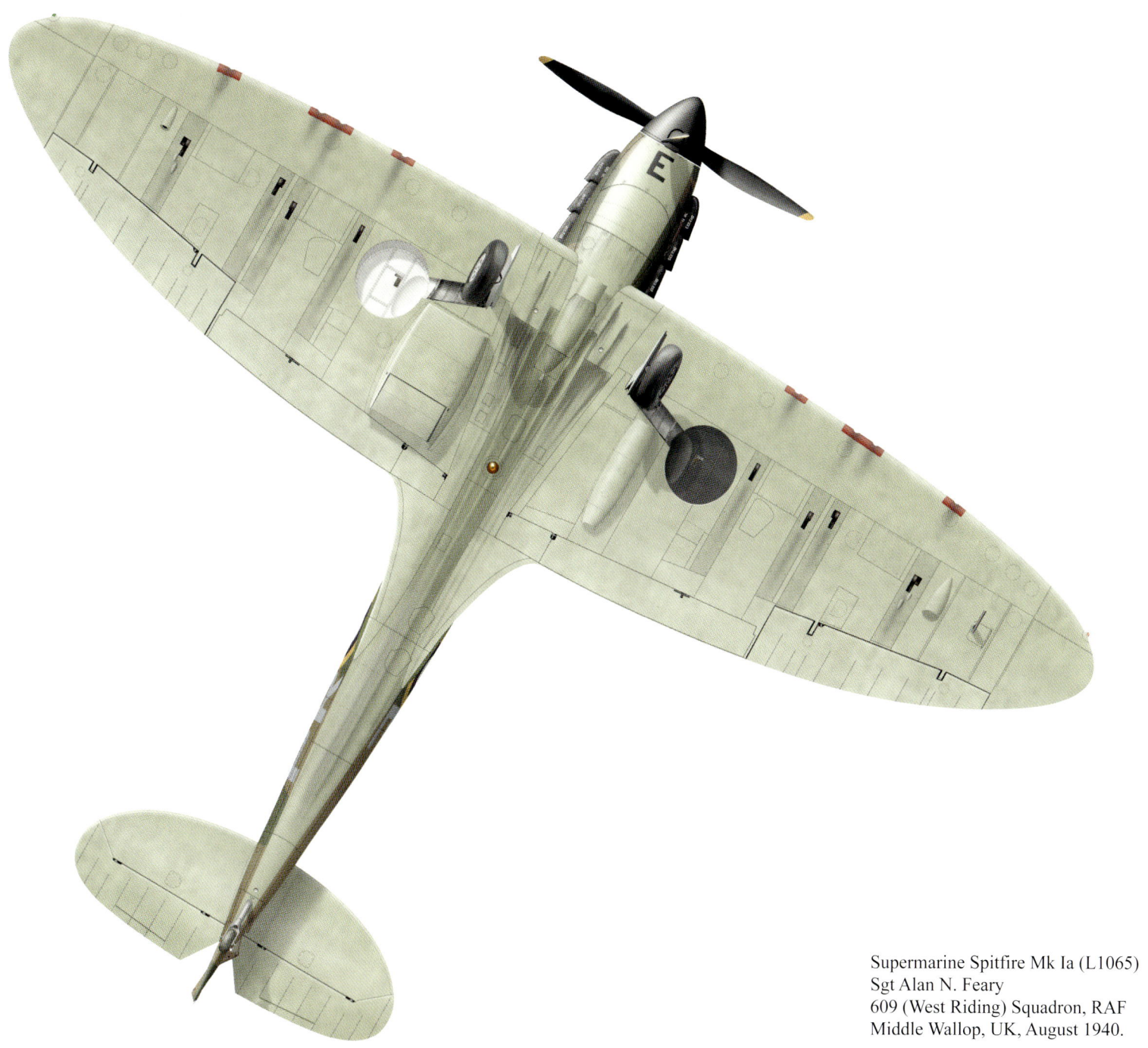

Supermarine Spitfire Mk Ia (L1065)
Sgt Alan N. Feary
609 (West Riding) Squadron, RAF
Middle Wallop, UK, August 1940.

Supermarine Spitfire Mk Ia (R6637)
Fg Off John Dundas
609 (West Riding) Squadron, RAF
Middle Wallop, UK, August 1940.

Supermarine Spitfire Mk Ia (N3277)
Plt Off Richard Hardy
234 Squadron, RAF
Middle Wallop, UK, August 1940.

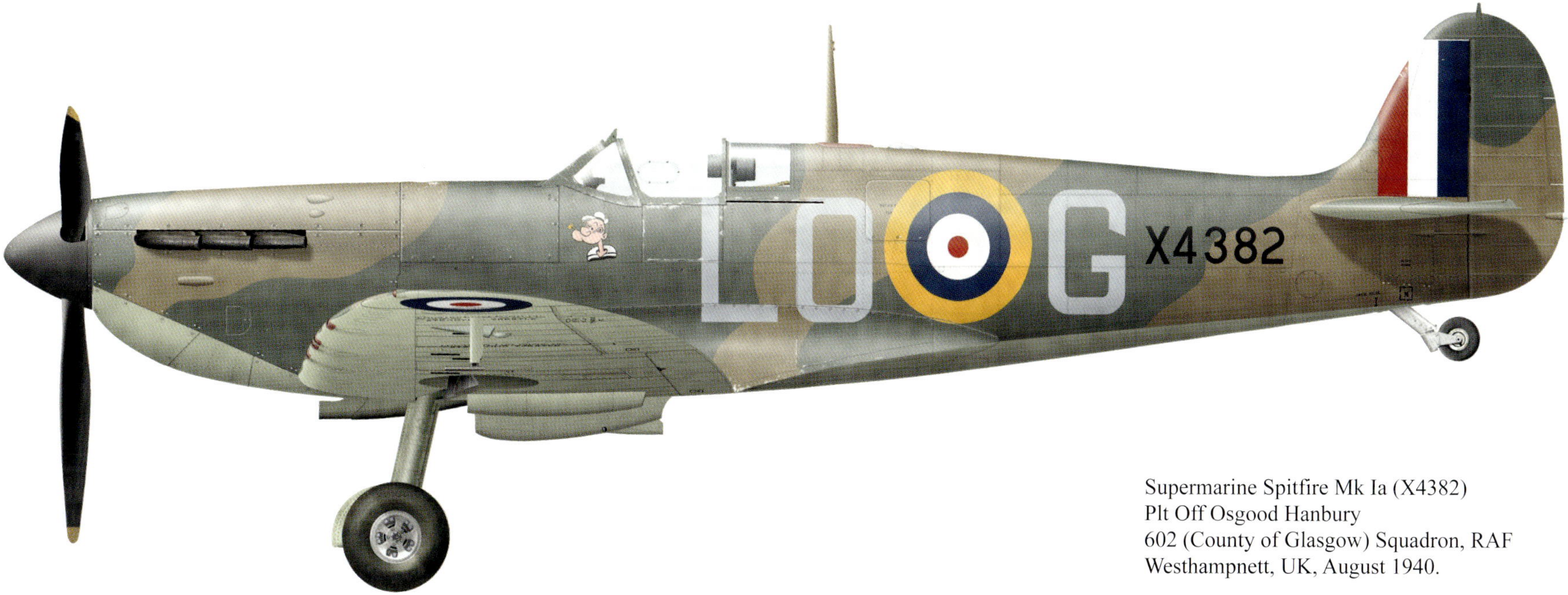

Supermarine Spitfire Mk Ia (X4382)
Plt Off Osgood Hanbury
602 (County of Glasgow) Squadron, RAF
Westhampnett, UK, August 1940.

Supermarine Spitfire Mk Ia (R6623)
Flt Sgt Jack Mann
64 Squadron, RAF
Kenley, UK, August 1940.

Supermarine Spitfire Mk Ia (P9386)
Sqn Ldr Brian 'Sandy' Lane
19 Squadron, RAF
Fowlmere, UK, September 1940.

Supermarine Spitfire Mk Ia (X4425)
Flt Sgt George 'Grumpy' Unwin
19 Squadron, RAF
Fowlmere, UK, September 1940.

Supermarine Spitfire Mk Ia (N3085)
Plt Off Hubert Allen
66 Squadron, RAF
Kenley, UK, September 1940.

Supermarine Spitfire Mk Ia (R6800)
Sqn Ldr Rupert Leigh
66 Squadron, RAF
Gravesend, UK, September 1940.

Supermarine Spitfire Mk Ia (R6883)
Flt Lt Gordon Olive
65 (East India) Squadron, RAF
Tangmere, UK, December 1940.

Supermarine Spitfire Mk Ib (R6908)
Flt Lt Brian Kingcome
92 (East India) Squadron, RAF
Biggin Hill, UK, December 1940.

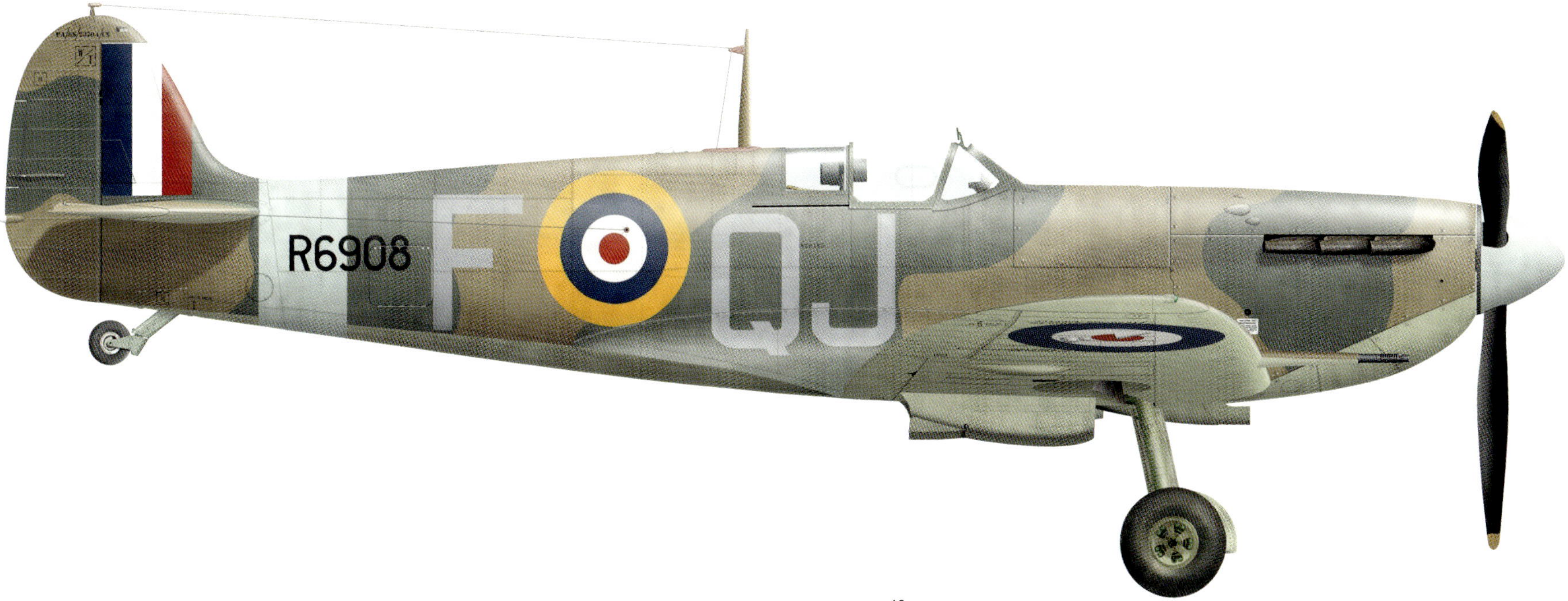

43

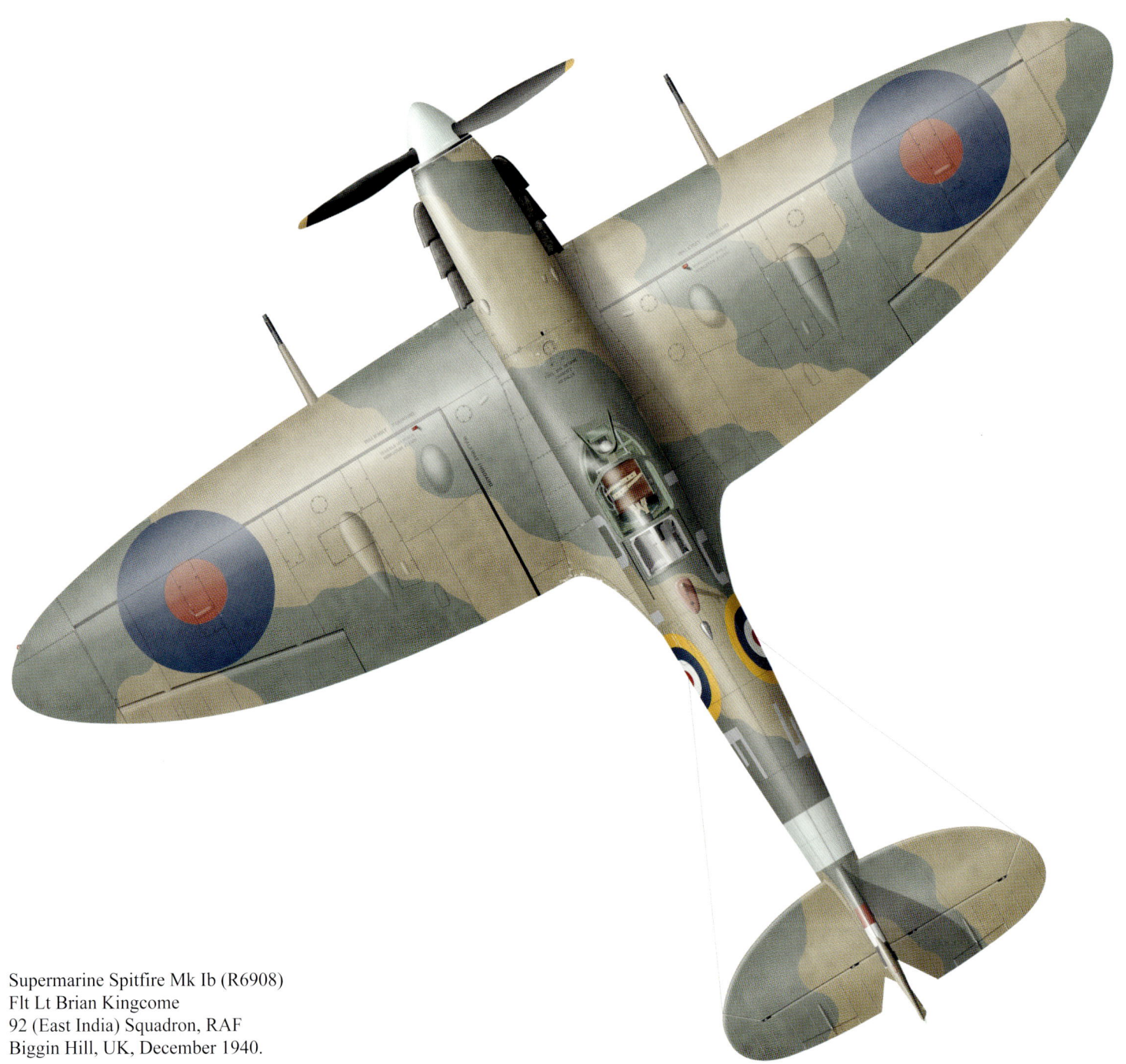

Supermarine Spitfire Mk Ib (R6908)
Flt Lt Brian Kingcome
92 (East India) Squadron, RAF
Biggin Hill, UK, December 1940.

Supermarine Spitfire Mk Ib (R6908)
Flt Lt Brian Kingcome
92 (East India) Squadron, RAF
Biggin Hill, UK, December 1940.

Supermarine Spitfire Mk Ia (X4253)
Plt Off Wilfrid Duncan Smith
611 (West Lancashire) Squadron, RAF
Hornchurch, UK, February 1941.

Supermarine Spitfire Mk Ia (X4828)
Flt Lt Wojciech Kolaczkowski, 303 (Polish) Squadron, RAF, Speke UK, September 1941.
In mid-August 1941, RAF Fighter Command introduced a new day fighter scheme by replacing the
upper surfaces Dark Earth areas with Dark Sea Grey, and the Sky under surfaces with Medium Sea Grey.

Spitfire Mk II

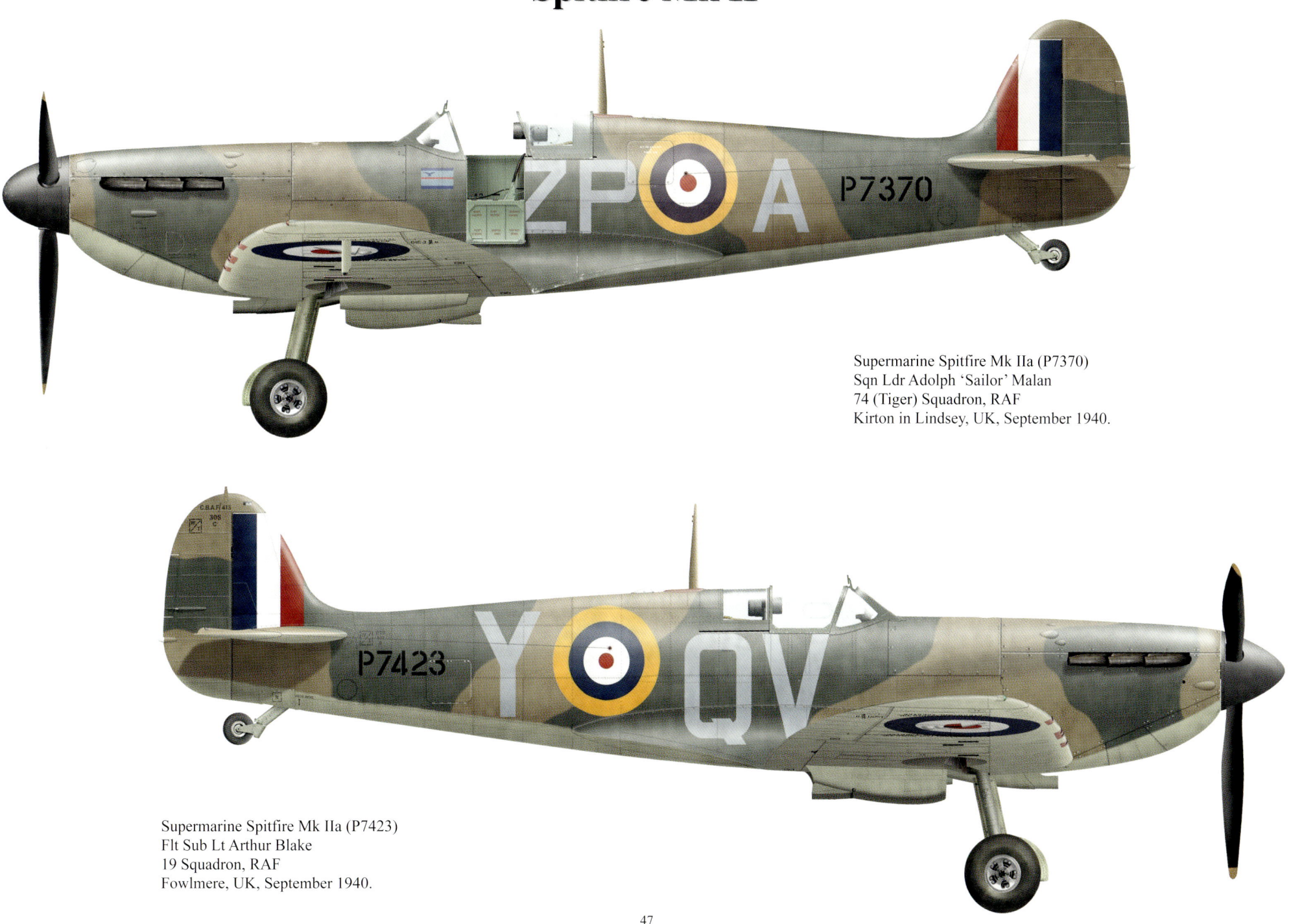

Supermarine Spitfire Mk IIa (P7370)
Sqn Ldr Adolph 'Sailor' Malan
74 (Tiger) Squadron, RAF
Kirton in Lindsey, UK, September 1940.

Supermarine Spitfire Mk IIa (P7423)
Flt Sub Lt Arthur Blake
19 Squadron, RAF
Fowlmere, UK, September 1940.

Supermarine Spitfire Mk IIa (P7666)
Sqn Ldr Donald Finlay
41 Squadron, RAF
Hornchurch, UK, December 1940.

Supermarine Spitfire Mk IIa (P7916)
Sgt William Johnson
145 Squadron, RAF
Tangmere, UK, spring 1941.

Supermarine Spitfire Mk IIa (P7881)
Sqn Ldr Michael Robinson
609 (West Riding) Squadron, RAF
Biggin Hill, UK, April 1941.

Supermarine Spitfire Mk IIa (P7881)
Sqn Ldr Michael Robinson
609 (West Riding) Squadron, RAF
Biggin Hill, UK, April 1941.

Supermarine Spitfire Mk IIa (P7881)
Sqn Ldr Michael Robinson
609 (West Riding) Squadron, RAF
Biggin Hill, UK, April 1941.

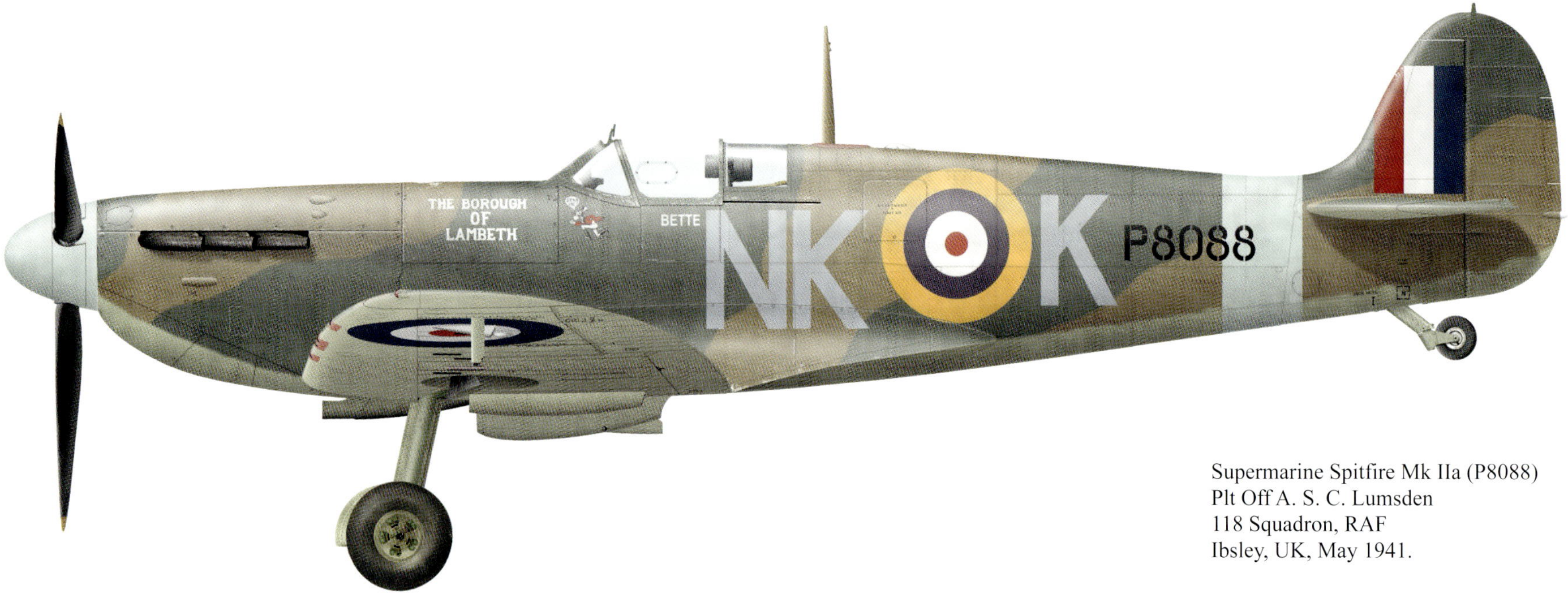

Supermarine Spitfire Mk IIa (P8088)
Plt Off A. S. C. Lumsden
118 Squadron, RAF
Ibsley, UK, May 1941.

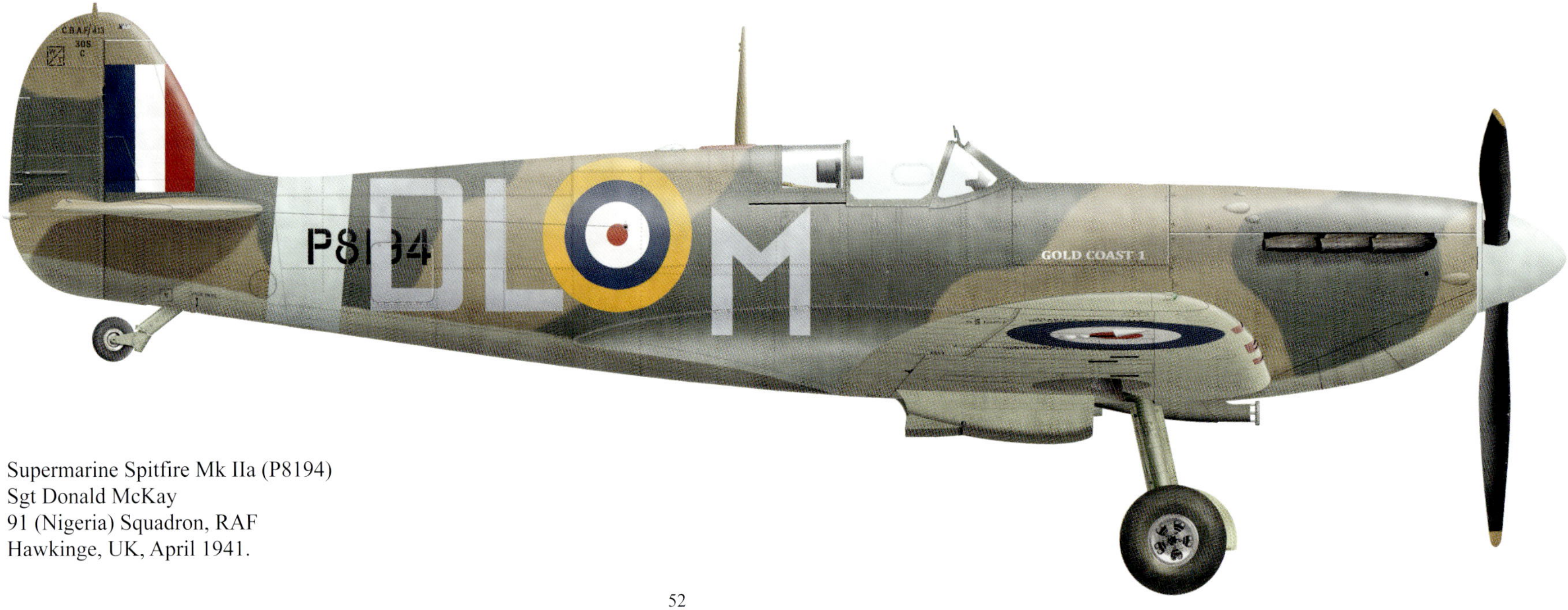

Supermarine Spitfire Mk IIa (P8194)
Sgt Donald McKay
91 (Nigeria) Squadron, RAF
Hawkinge, UK, April 1941.

Supermarine Spitfire Mk IIb (P8385)
Fg Off Miroslav Feric
303 (Polish) Squadron, RAF
Northolt, UK, June 1941.

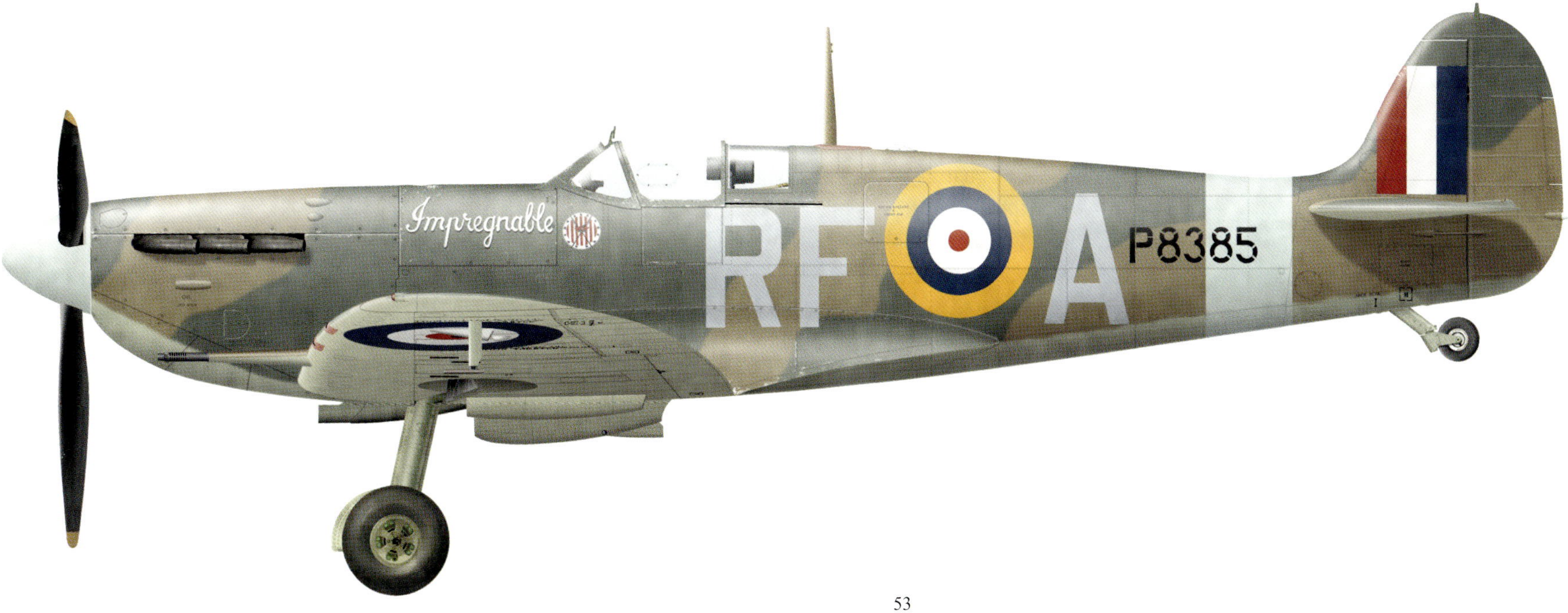

Supermarine Spitfire Mk IIa (P7855)
Fg Off Jan Falkowski
315 (Polish) Squadron, RAF
Northolt, UK, summer 1941.

Supermarine Spitfire Mk IIa (P8038)
Flt Lt Brendan 'Paddy' Finucane
452 Squadron, RAAF
Kenley, UK, August 1941.

Supermarine Spitfire Mk IIa (P7308)
Plt Off William 'Pappy' Dunn
71 (Eagle) Squadron, RAF
North Weald, UK, August 1941.

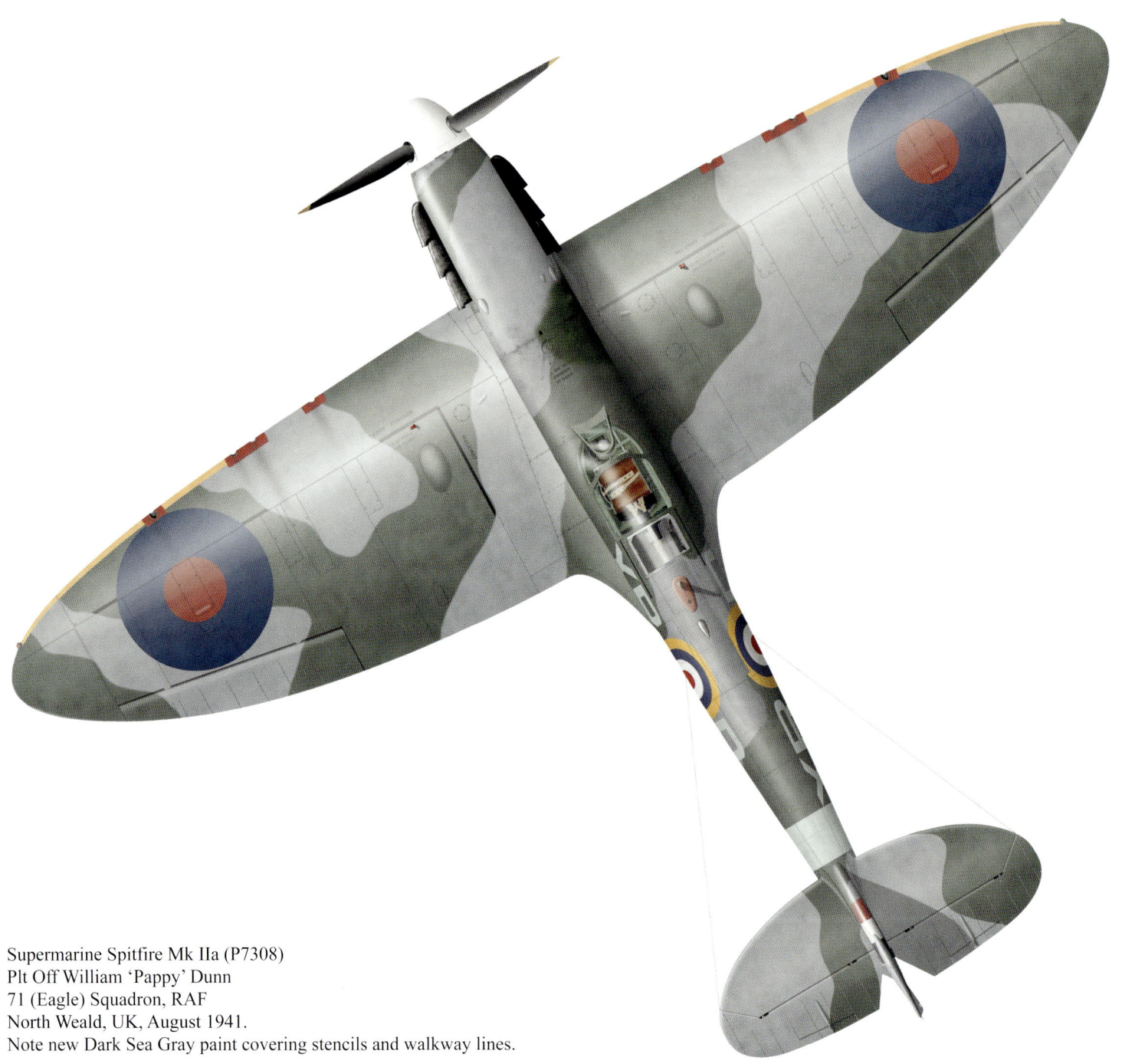

Supermarine Spitfire Mk IIa (P7308)
Plt Off William 'Pappy' Dunn
71 (Eagle) Squadron, RAF
North Weald, UK, August 1941.
Note new Dark Sea Gray paint covering stencils and walkway lines.

Supermarine Spitfire Mk IIa (P7308)
Plt Off William 'Pappy' Dunn
71 (Eagle) Squadron, RAF
North Weald, UK, August 1941.

Supermarine Spitfire Mk IIb (P8342)
Sgt Marcin Machowiak
306 (Polish) Squadron, RAF
Northolt, UK, August 1941.

Supermarine Spitfire Mk IIb (P8342)
Sgt Marcin Machowiak
306 (Polish) Squadron, RAF
Northolt, UK, August 1941.

Supermarine Spitfire Mk IIb (P8505)
Sgt Eric Dicks-Sherwood
266 (Rhodesia) Squadron, RAF
Wittering, UK, September 1941.

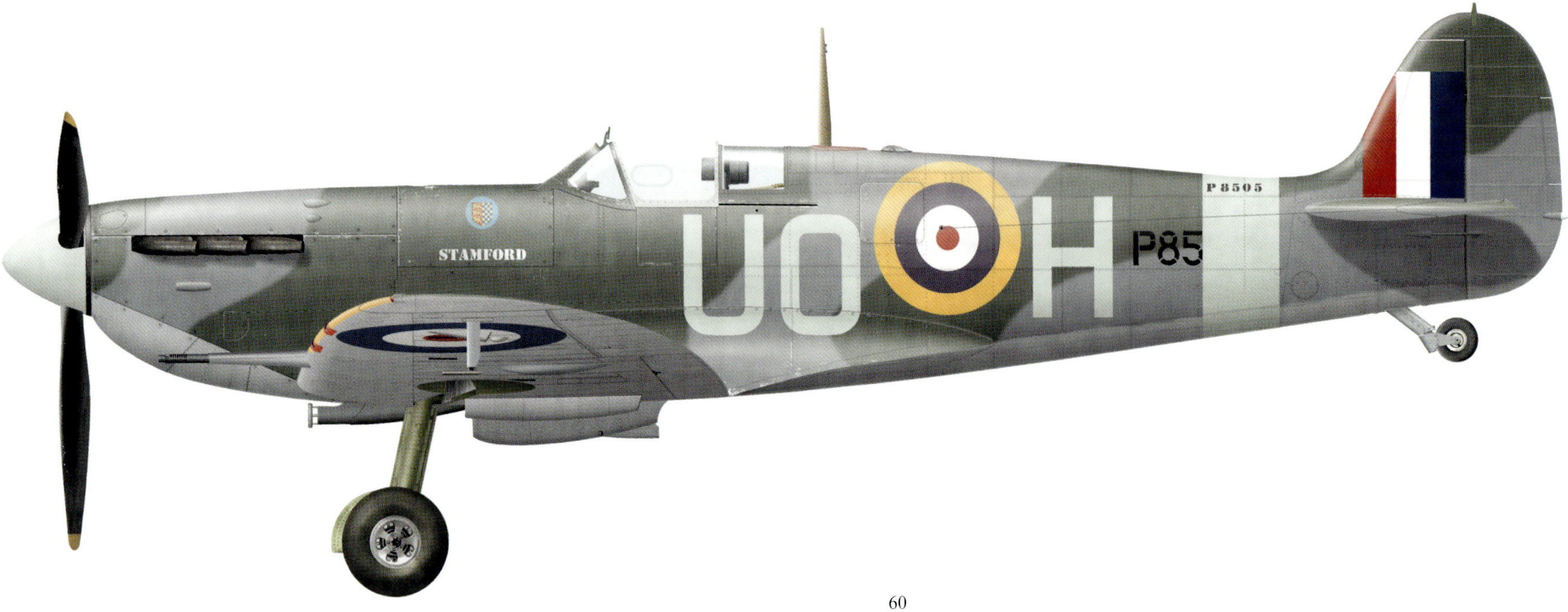

Supermarine Spitfire Mk IIa (P8397)
310 (Czechoslovak) Squadron, RAF
Dyce, UK, October 1941.

Supermarine Spitfire Mk IIa (P8084)
Flt Lt James O'Meara
64 Squadron, RAF
Drem, UK, October 1941.

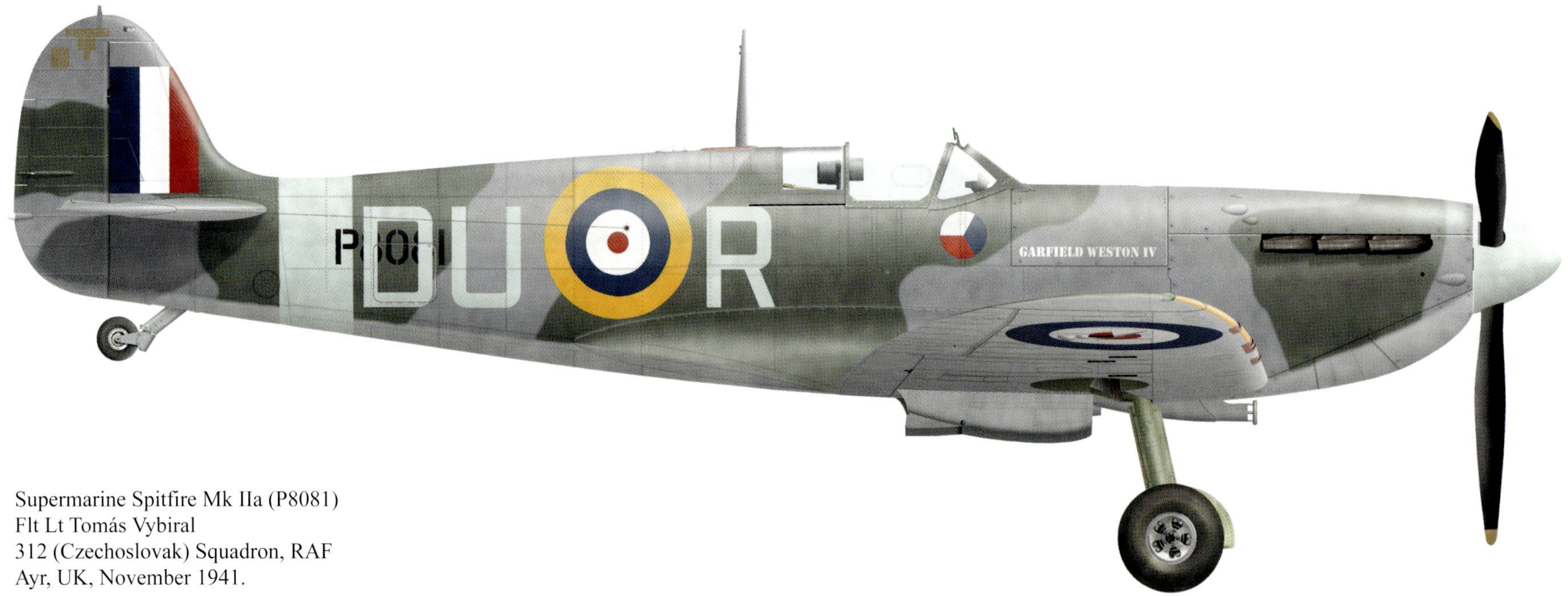

Supermarine Spitfire Mk IIa (P8081)
Flt Lt Tomás Vybiral
312 (Czechoslovak) Squadron, RAF
Ayr, UK, November 1941.

Supermarine Spitfire Mk IIa (P7840)
340 (Free French) Squadron, RAF
Turnhouse, UK, January 1942.

Spitfire Mk V

Supermarine Spitfire Mk Va (R7208)
611 (West Lancashire) Squadron, RAF
Hornchurch, UK, summer 1941.

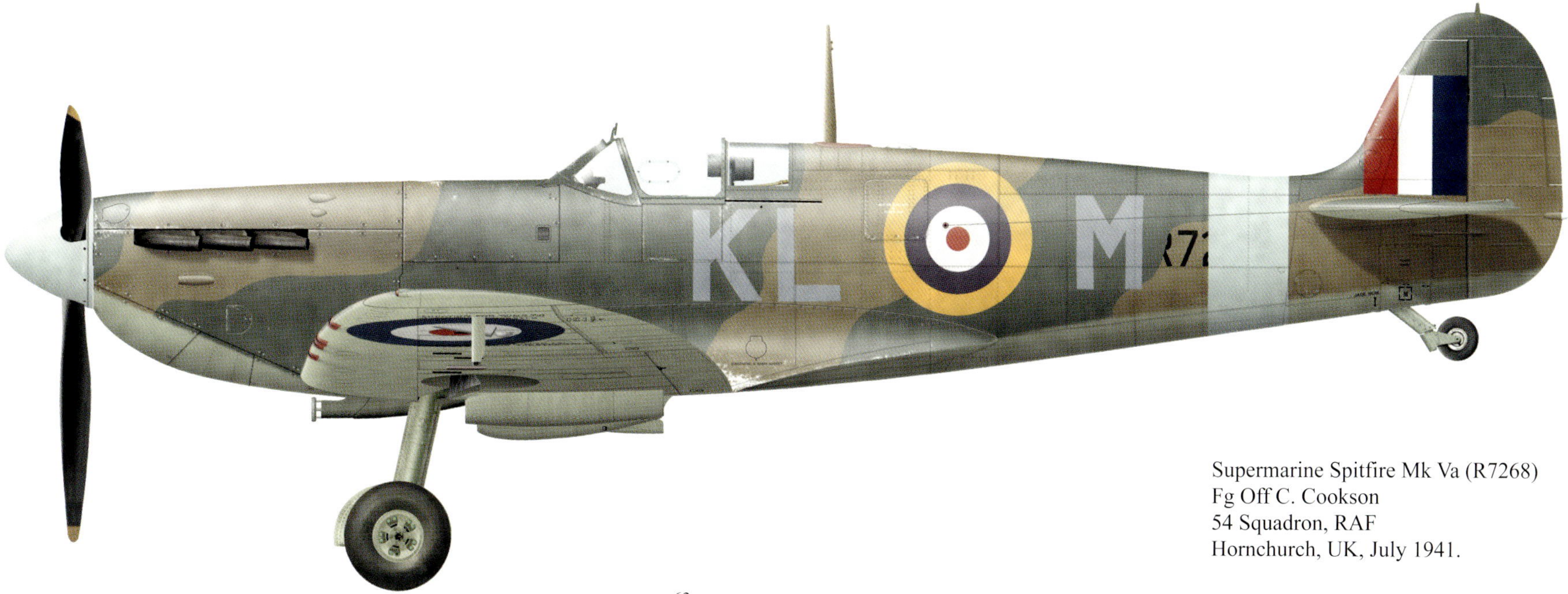

Supermarine Spitfire Mk Va (R7268)
Fg Off C. Cookson
54 Squadron, RAF
Hornchurch, UK, July 1941.

Supermarine Spitfire Mk Vb (W3312)
Sqn Ldr James Rankin
92 (East India) Squadron, RAF
Biggin Hill, UK, August 1941.

Supermarine Spitfire Mk Vb (W3122)
Flt Lt Jean Demozay
91 (Nigeria) Squadron, RAF
Hawkinge, UK, August 1941.

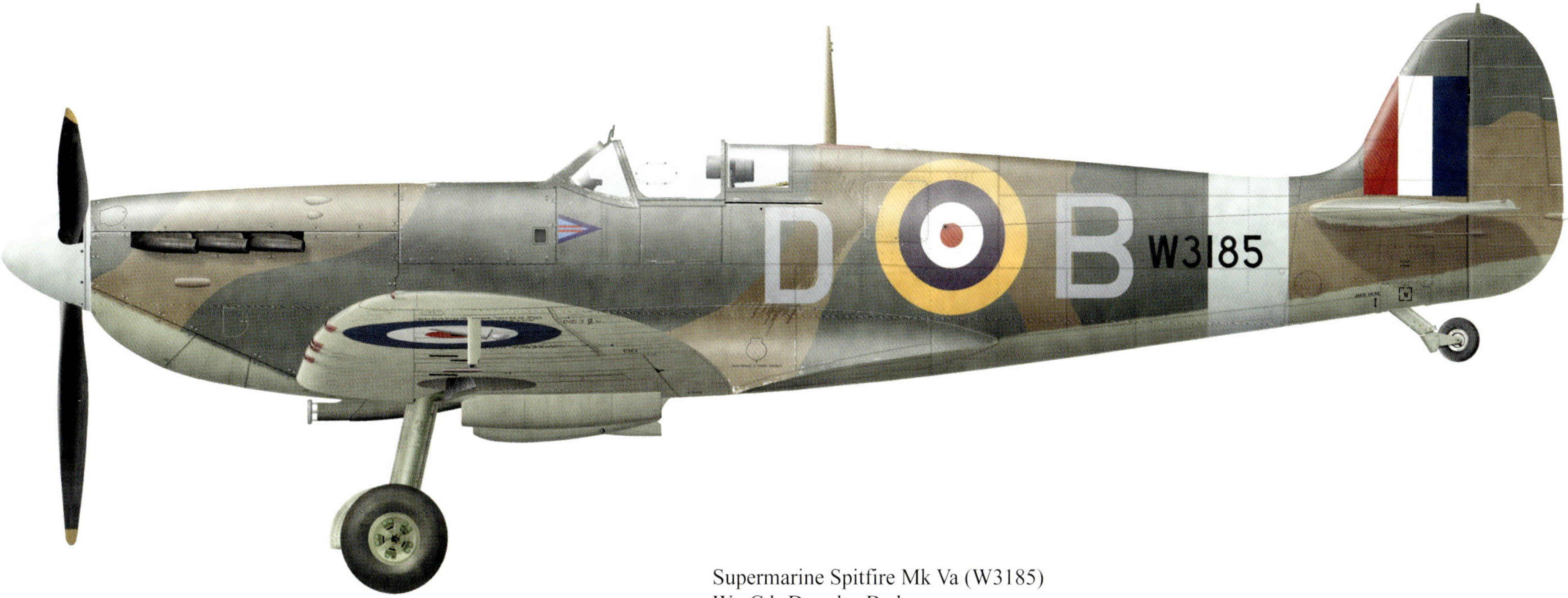

Supermarine Spitfire Mk Va (W3185)
Wg Cdr Douglas Bader
Tangmere Wing, RAF
Tangmere, UK, August 1941.

Supermarine Spitfire Mk Va (W3185)
Wg Cdr Douglas Bader
Tangmere Wing, RAF
Tangmere, UK, August 1941.

Supermarine Spitfire Mk Va (W3185)
Wg Cdr Douglas Bader
Tangmere Wing, RAF
Tangmere, UK, August 1941.

Supermarine Spitfire Mk Vb (W3507)
Plt Off James 'Jimmy' Whalen
129 Squadron, RAF
Tangmere, UK, September 1941.

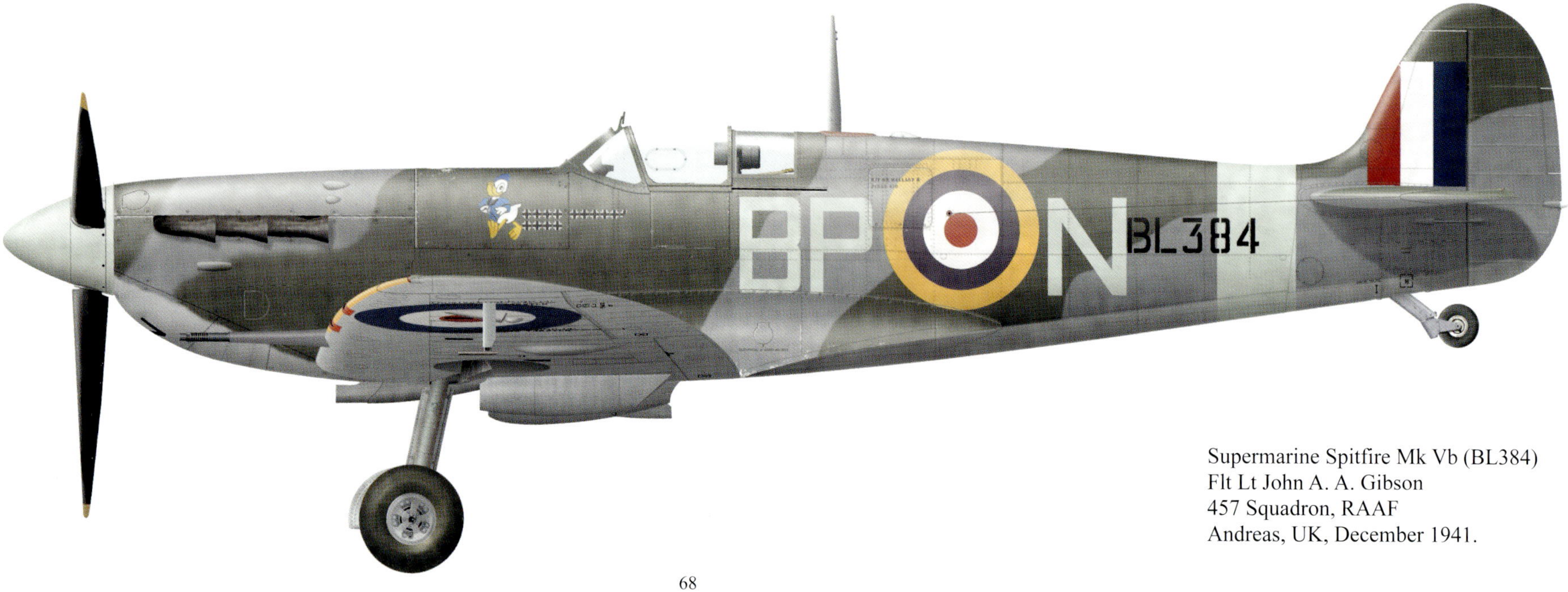

Supermarine Spitfire Mk Vb (BL384)
Flt Lt John A. A. Gibson
457 Squadron, RAAF
Andreas, UK, December 1941.

Supermarine Spitfire Mk Vb/Trop (AB262)
Flt Lt P. W. E. Heppell
249 Squadron, RAF
Takali, Malta, March 1942.

Supermarine Spitfire Mk Vb/Trop (AB264)
Plt Off Peter Nash, 249 Squadron, RAF, Takali, Malta, March 1942.
The desert camouflage of the tropicalised Spitfires that operated in the defence of Malta was modified with colours more suitable for fighting over the sea such as Mediterranean Blue or USN Blue Grey.

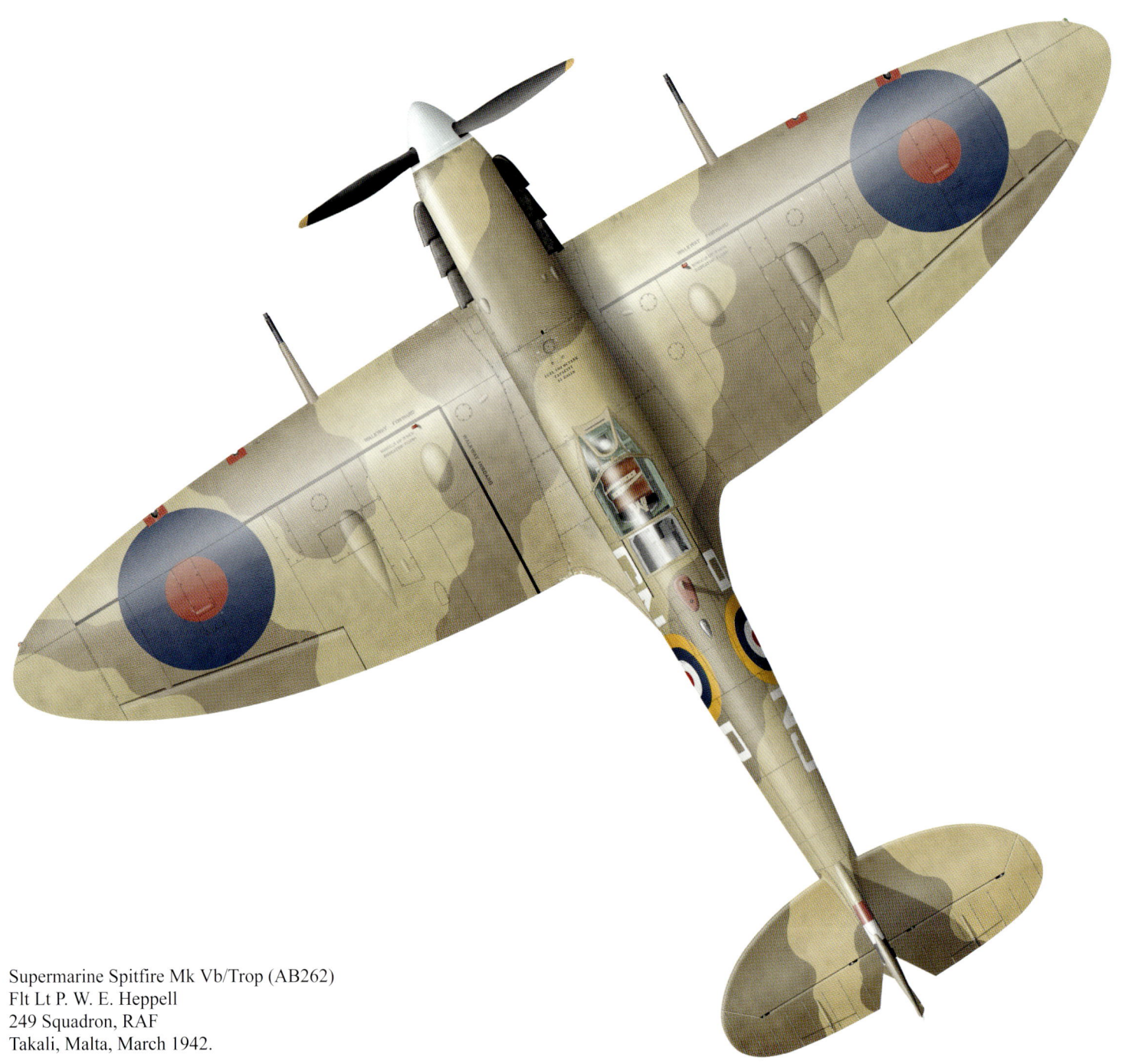

Supermarine Spitfire Mk Vb/Trop (AB262)
Flt Lt P. W. E. Heppell
249 Squadron, RAF
Takali, Malta, March 1942.

Supermarine Spitfire Mk Vb/Trop (AB262)
Flt Lt P. W. E. Heppell
249 Squadron, RAF
Takali, Malta, March 1942.

Supermarine Spitfire Mk Vb (AA865)
Sgt Václav Truháf
313 (Czechoslovak) Squadron, RAF
Hornchurch, UK, April 1942.

Supermarine Spitfire Mk Vb (BL753)
Plt Off Donald 'Don' Blakeslee
401 Squadron, RCAF
Gravesend, UK, April 1942.

Supermarine Spitfire Mk Vc (AB380)
Wg Cdr Ian Gleed
Ibsley Wing, RAF
Ibsley, UK, April 1942.

Supermarine Spitfire Mk Vb (BM263)
Sqn Ldr Eric H. Thomas
133 (Eagle) Squadron, RAF
Kirton in Lindsey, UK, spring 1942.

Supermarine Spitfire Mk Vc/Trop (BR124)
603 (City of Edinburgh) Squadron, RAF
Takali, Malta, April 1942.

Supermarine Spitfire Mk Vc/Trop (BR349)
Plt Off John L. Boyd
185 Squadron, RAF
Takali, Malta, May 1942.

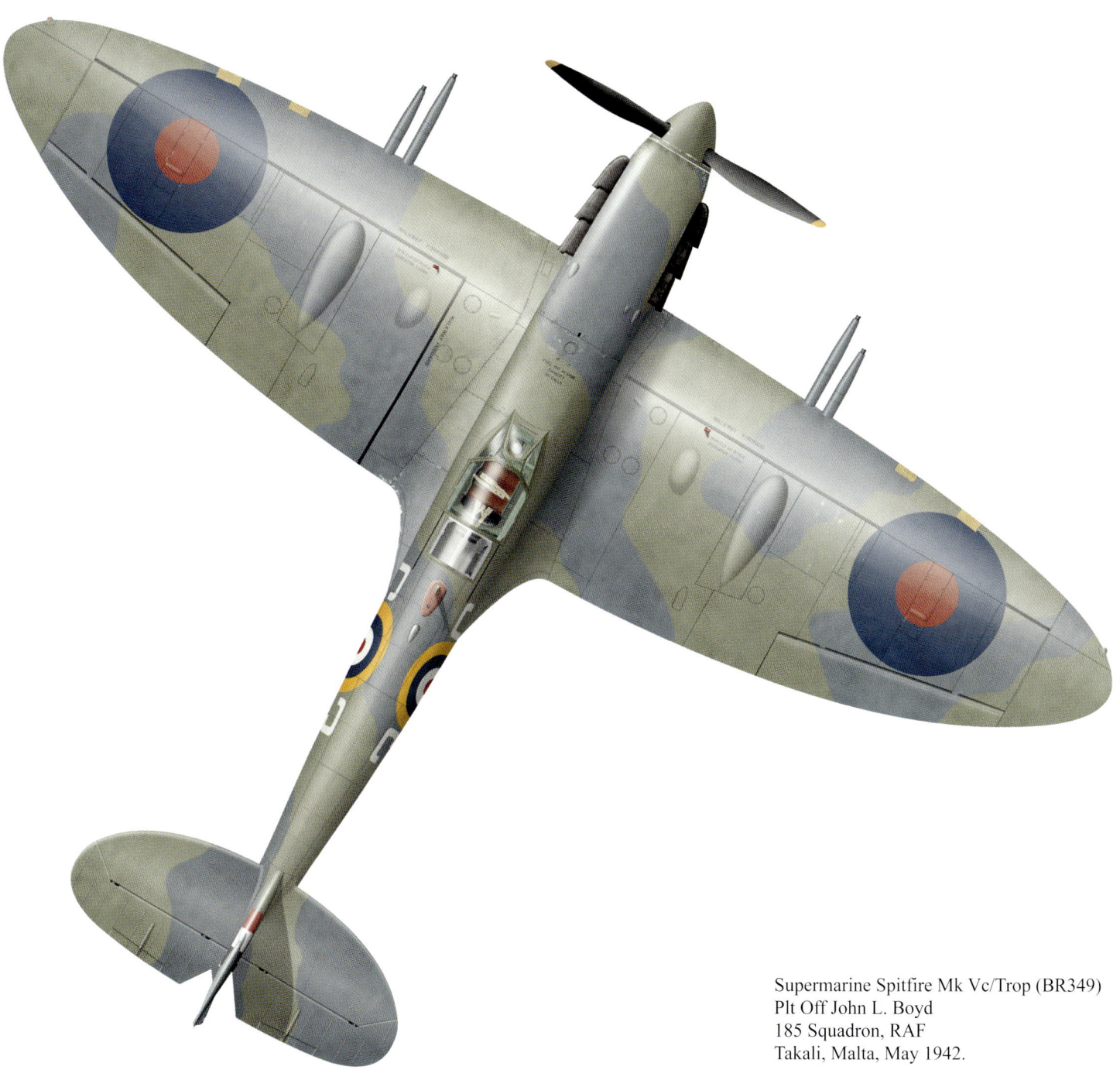

Supermarine Spitfire Mk Vc/Trop (BR349)
Plt Off John L. Boyd
185 Squadron, RAF
Takali, Malta, May 1942.

Supermarine Spitfire Mk Vb (BM144)
Flt Lt Jan Zumbach
303 (Polish) Squadron, RAF
Northolt, UK, May 1942.

Supermarine Spitfire Mk Vb (BL891)
332 (Norwegian) Squadron, RAF
North Weald, UK, May 1942.

Supermarine Spitfire Mk Vc/Trop (BR126)
Plt Off Jerry Smith
126 Squadron, RAF
Luqa, Malta, May 1942.

Supermarine Spitfire Mk Vc/Trop (BR246)
Plt Off Frank Jones
249 Squadron, RAF
Takali, Malta, June 1942.

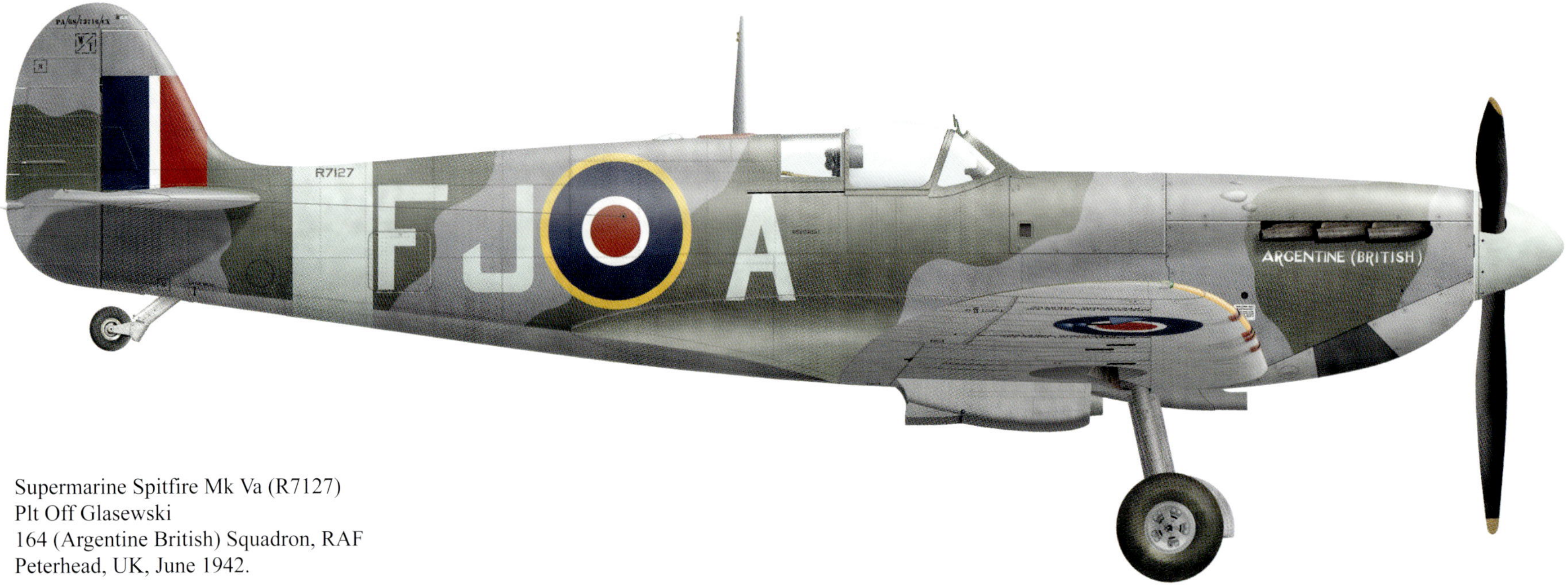

Supermarine Spitfire Mk Va (R7127)
Plt Off Glasewski
164 (Argentine British) Squadron, RAF
Peterhead, UK, June 1942.

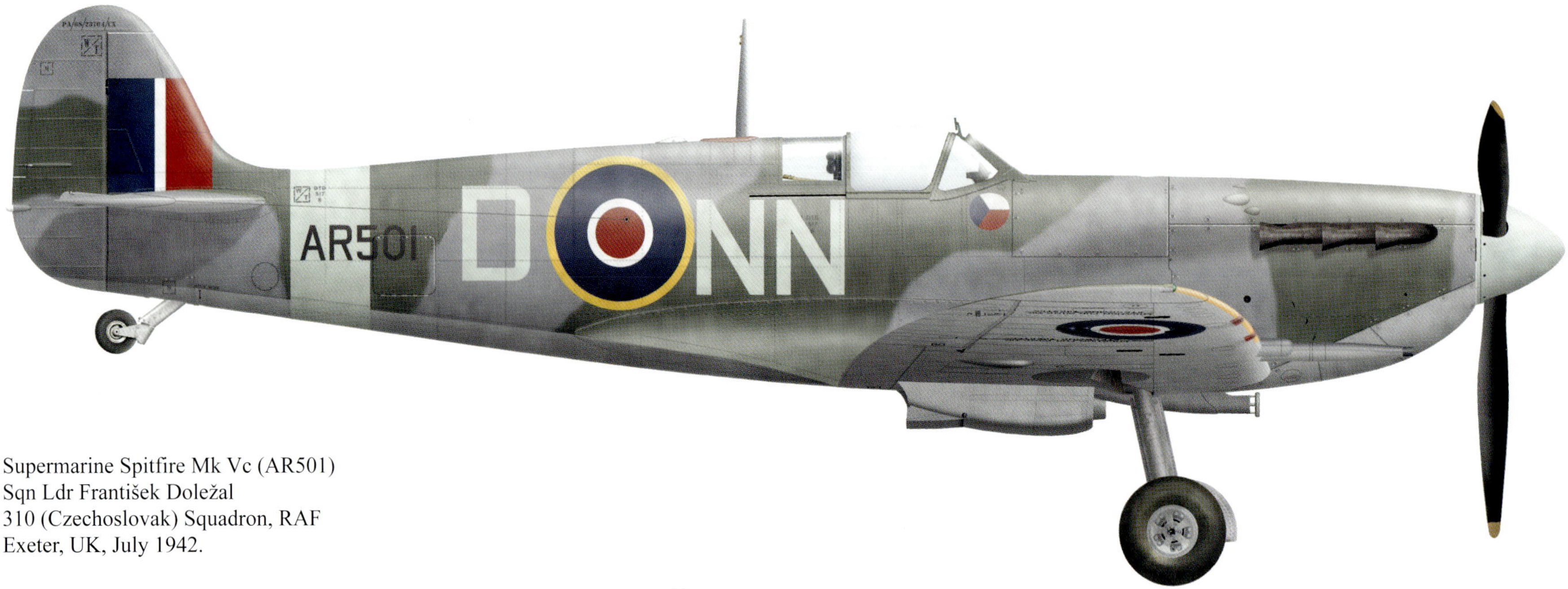

Supermarine Spitfire Mk Vc (AR501)
Sqn Ldr František Doležal
310 (Czechoslovak) Squadron, RAF
Exeter, UK, July 1942.

Supermarine Spitfire Mk Vb (BM324)
Wg Cdr Bernard Duperier
340 (Free French) Squadron, RAF
Hornchurch, UK, July 1942.

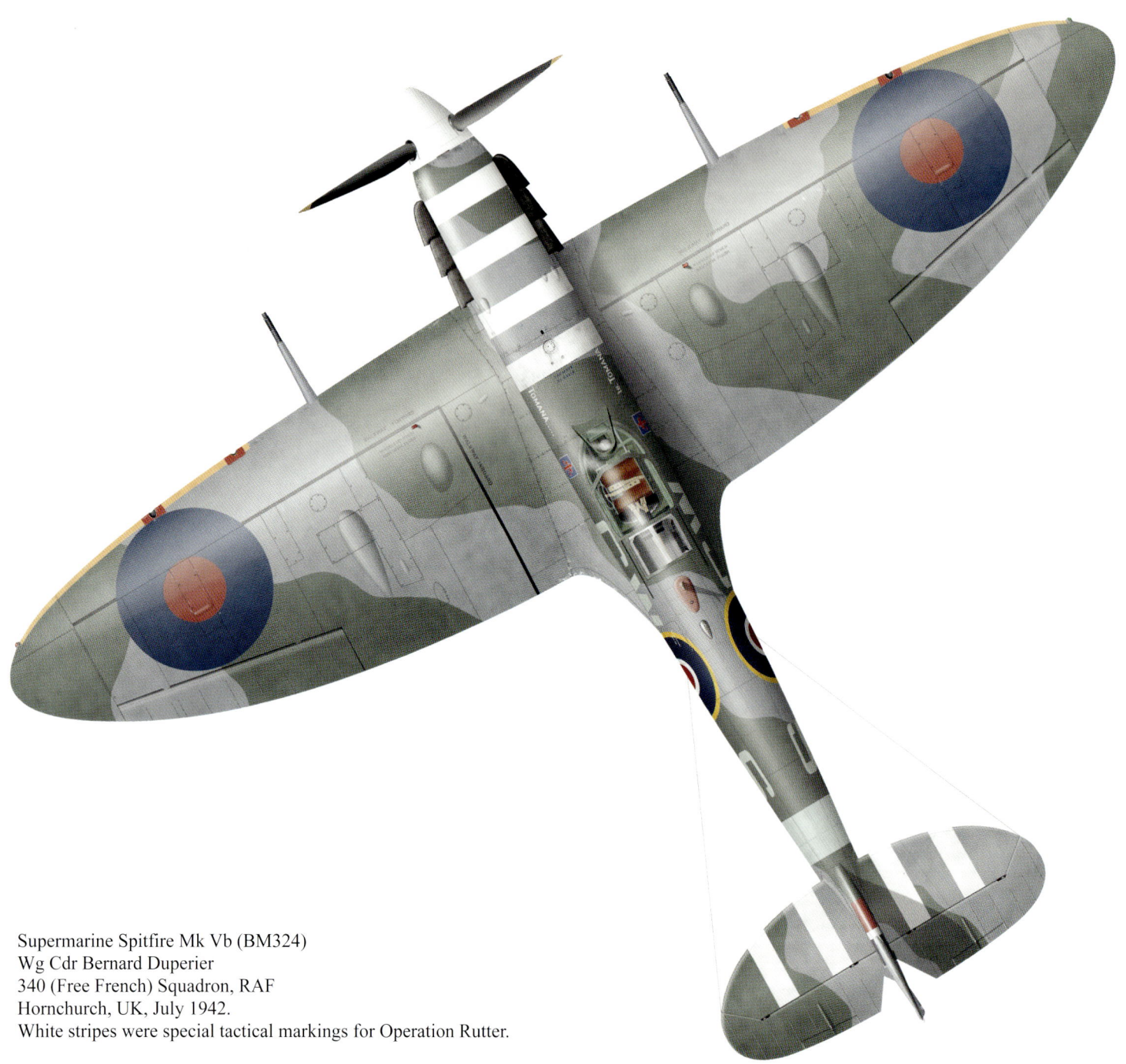

Supermarine Spitfire Mk Vb (BM324)
Wg Cdr Bernard Duperier
340 (Free French) Squadron, RAF
Hornchurch, UK, July 1942.
White stripes were special tactical markings for Operation Rutter.

Supermarine Spitfire Mk Vb (BM324)
Wg Cdr Bernard Duperier
340 (Free French) Squadron, RAF
Hornchurch, UK, July 1942.

Supermarine Spitfire Mk Vc/Trop (BR130)
Sgt George 'Screwball' Beurling
249 Squadron, RAF
Luqa, Malta, July 1942.

Supermarine Spitfire Mk Vc/Trop (BR323)
Sgt George 'Screwball' Beurlung
249 Squadron, RAF
Luqa, Malta, July 1942.

Supermarine Spitfire Mk Vc/Trop (BR387)
Plt Off John 'Slim' Yarra
185 Squadron, RAF
Hal Far, Malta, July 1942.

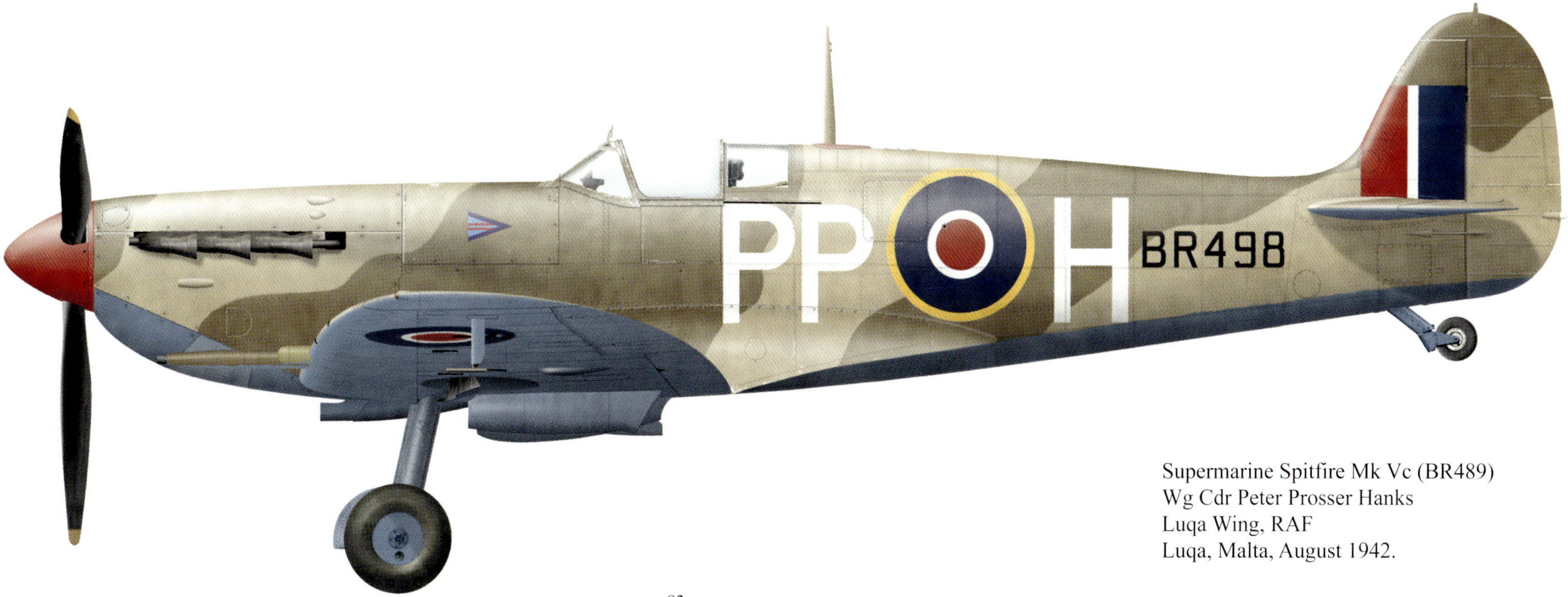

Supermarine Spitfire Mk Vc (BR489)
Wg Cdr Peter Prosser Hanks
Luqa Wing, RAF
Luqa, Malta, August 1942.

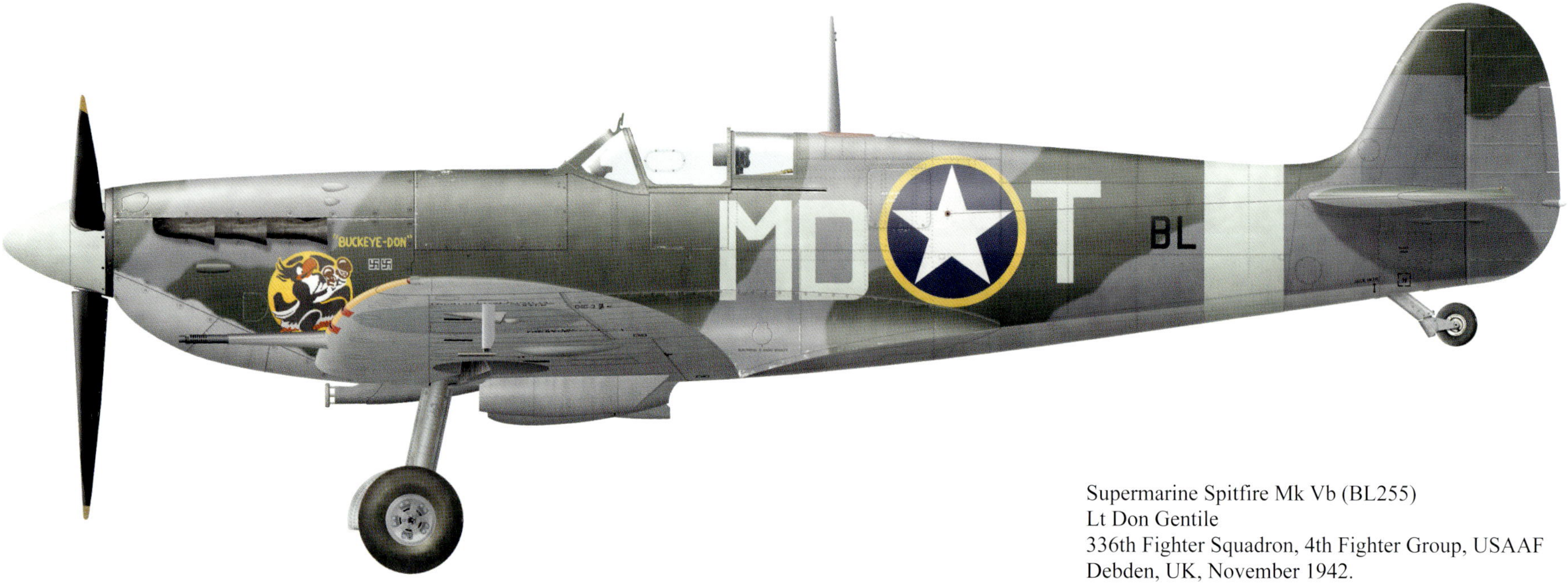

Supermarine Spitfire Mk Vb (BL255)
Lt Don Gentile
336th Fighter Squadron, 4th Fighter Group, USAAF
Debden, UK, November 1942.

Supermarine Spitfire Mk Vb (BM309)
2/Lt Robert A. Boock
335th Fighter Squadron, 4th Fighter Group, USAAF
Debden, UK, January 1943.

Supermarine Spitfire Mk Vc/Trop (A58-84)
Fg Off Frank 'Bush' Hamilton
457 Squadron, RAAF
Darwin, Australia, early 1943.

Supermarine Spitfire Mk Vb/Trop (ER821)
Flt Lt Neville Duke
92 (East India) Squadron, RAF
Bou Grara, Tunisia, January 1943.

Supermarine Spitfire Mk Vc (AR556)
1435 Squadron, RAF
Luqa, Malta, March 1943.

Supermarine Spitfire Mk Vb (EP829)
Sqn Ldr John J. Lynch
249 Squadron, RAF
Krendi, Malta, April 1943.

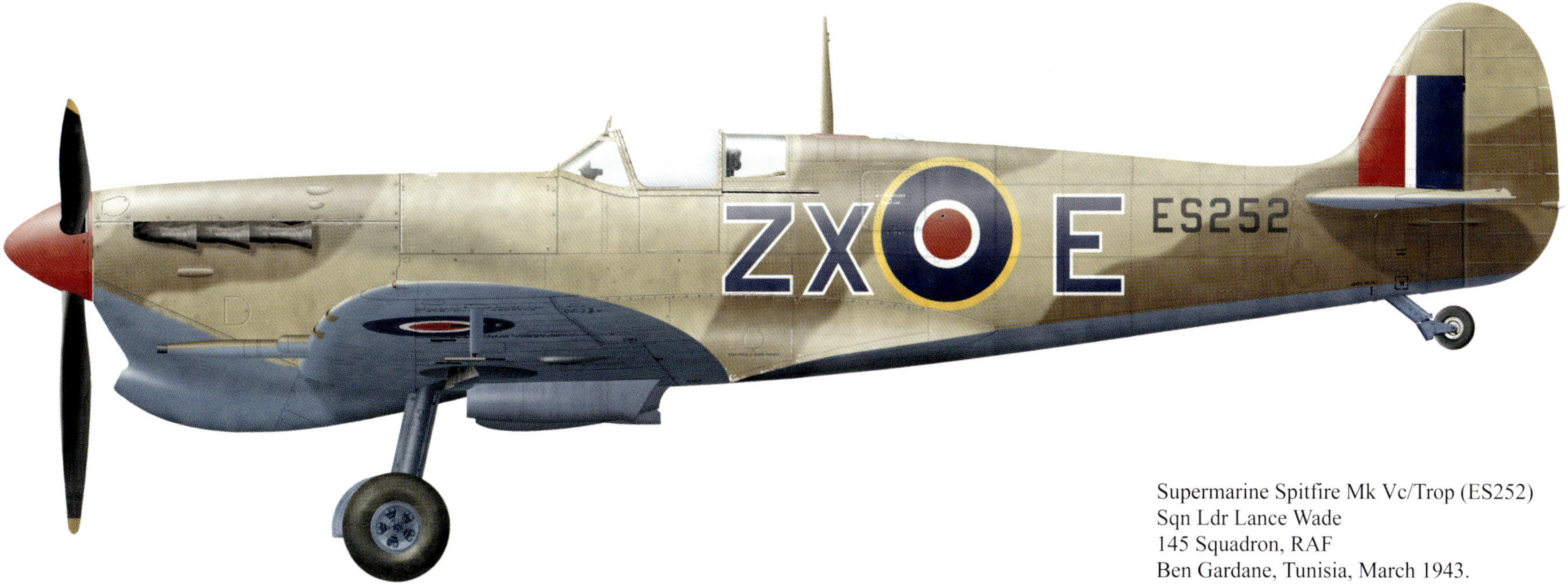

Supermarine Spitfire Mk Vc/Trop (ES252)
Sqn Ldr Lance Wade
145 Squadron, RAF
Ben Gardane, Tunisia, March 1943.

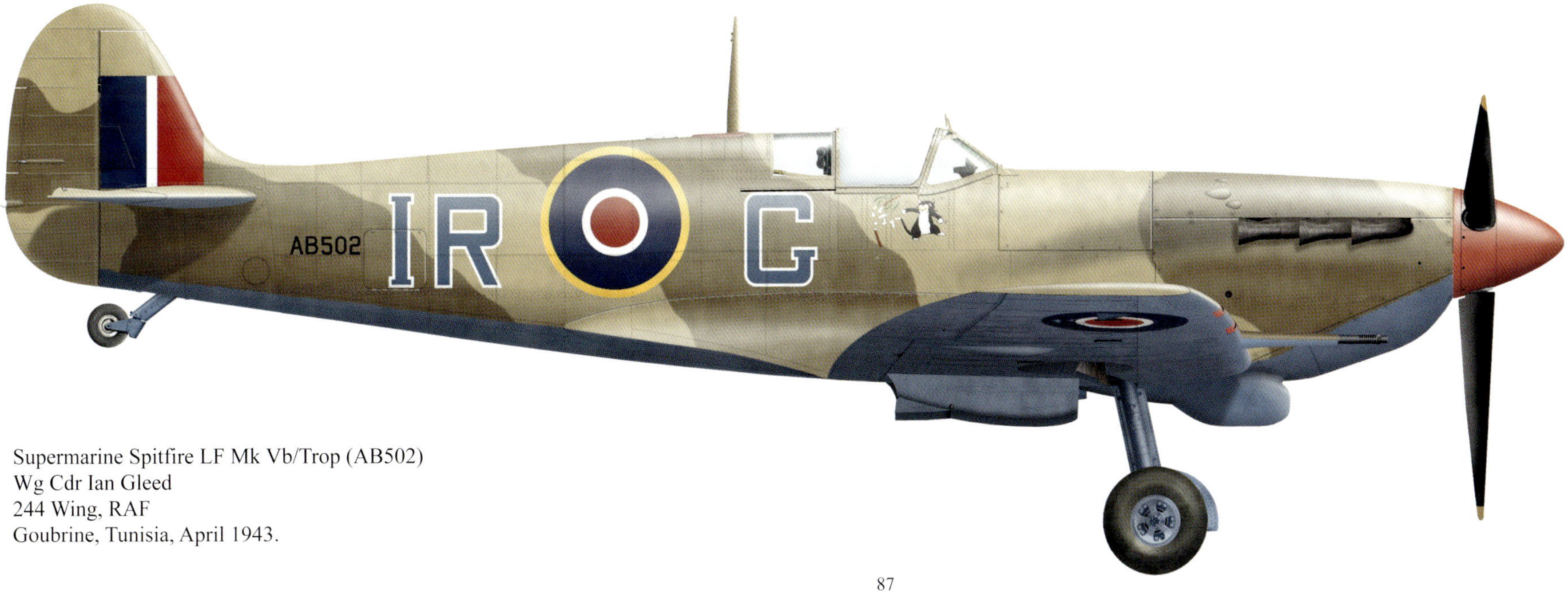

Supermarine Spitfire LF Mk Vb/Trop (AB502)
Wg Cdr Ian Gleed
244 Wing, RAF
Goubrine, Tunisia, April 1943.

Supermarine Spitfire Mk Vb (EP210)
57 GvIAP, VVS-RKKA
Kuban, USSR, April 1943.

Supermarine Spitfire LF Mk Vb (EN921)
Fg Off Jack Sheppard
401 (City of Westmount) Squadron, RCAF
Redhill, UK, July 1943.

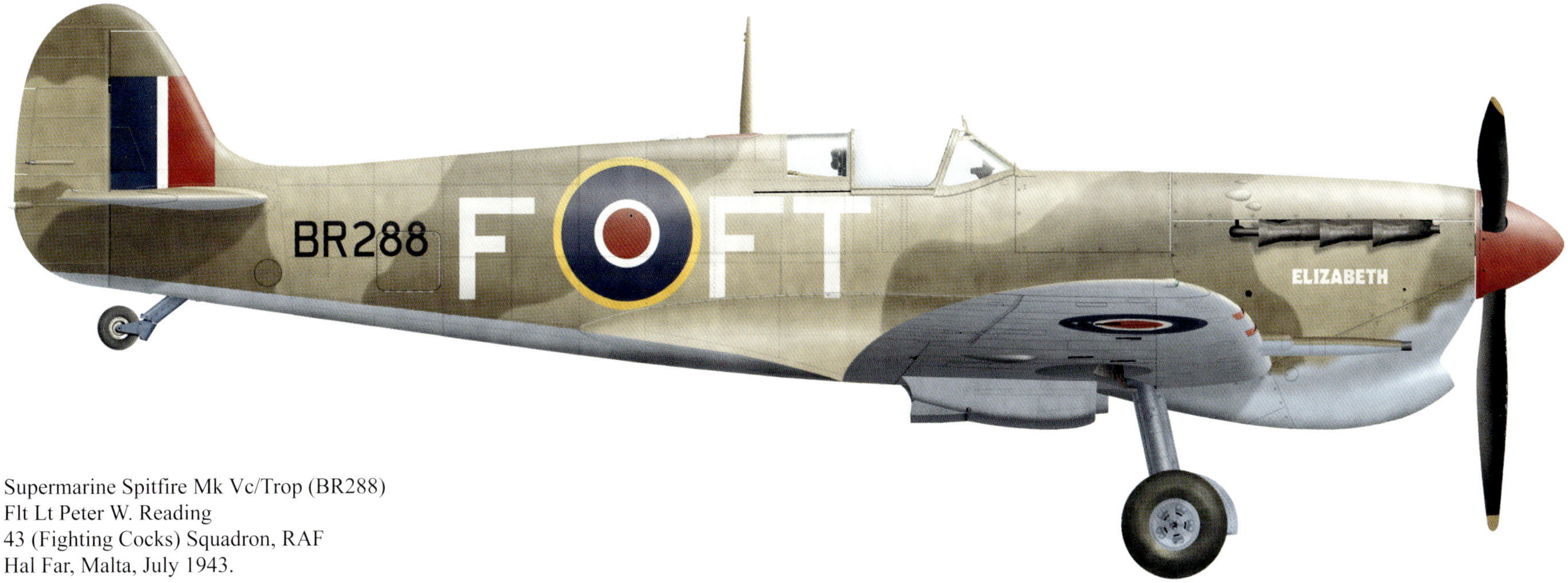

Supermarine Spitfire Mk Vc/Trop (BR288)
Flt Lt Peter W. Reading
43 (Fighting Cocks) Squadron, RAF
Hal Far, Malta, July 1943.

Supermarine Spitfire LF Mk Vb/Trop (EP689)
Sqn Ldr Stanislaw Skalski
601 (County of London) Squadron, RAF
Pachino, Italy, July 1943.

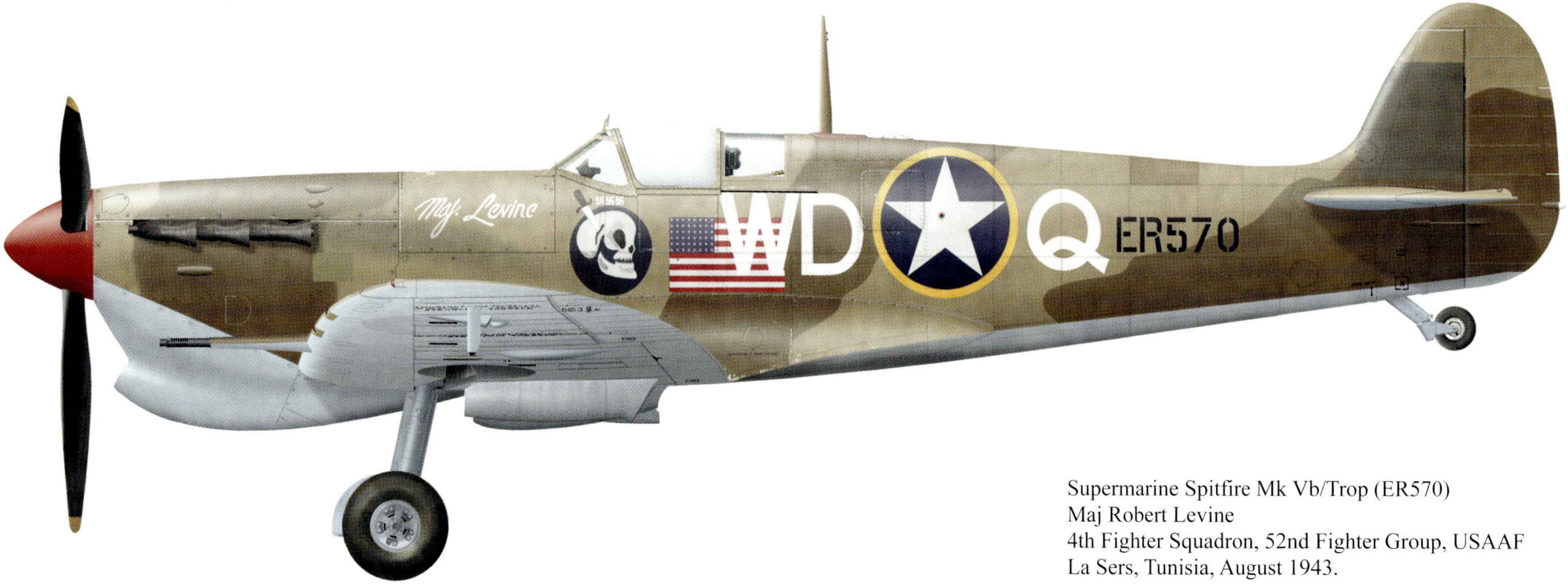

Supermarine Spitfire Mk Vb/Trop (ER570)
Maj Robert Levine
4th Fighter Squadron, 52nd Fighter Group, USAAF
La Sers, Tunisia, August 1943.

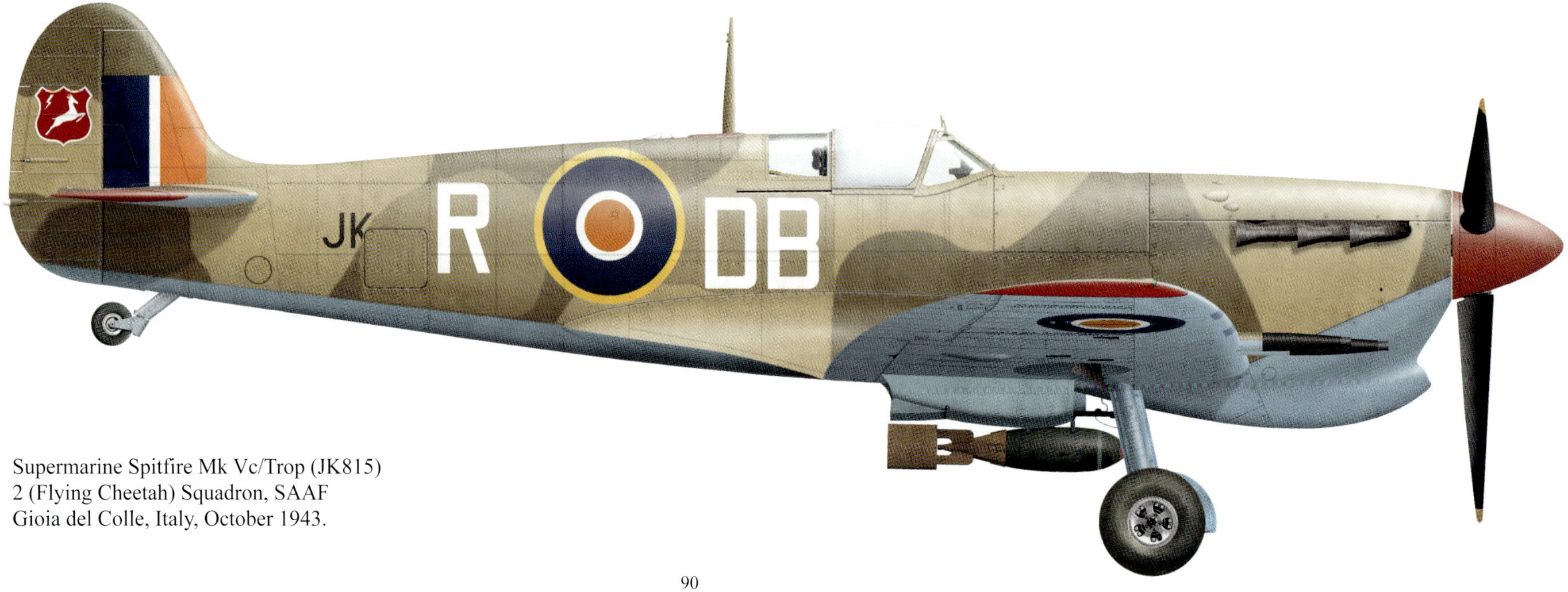

Supermarine Spitfire Mk Vc/Trop (JK815)
2 (Flying Cheetah) Squadron, SAAF
Gioia del Colle, Italy, October 1943.

Supermarine Spitfire Mk Vc/Trop (JK815)
2 (Flying Cheetah) Squadron, SAAF
Gioia del Colle, Italy, October 1943.

Supermarine Spitfire Mk Vc/Trop (A58-137)
Flt Lt David Hopton
79 Squadron, RAAF
Kiriwina, Trobriand Is, October 1943.

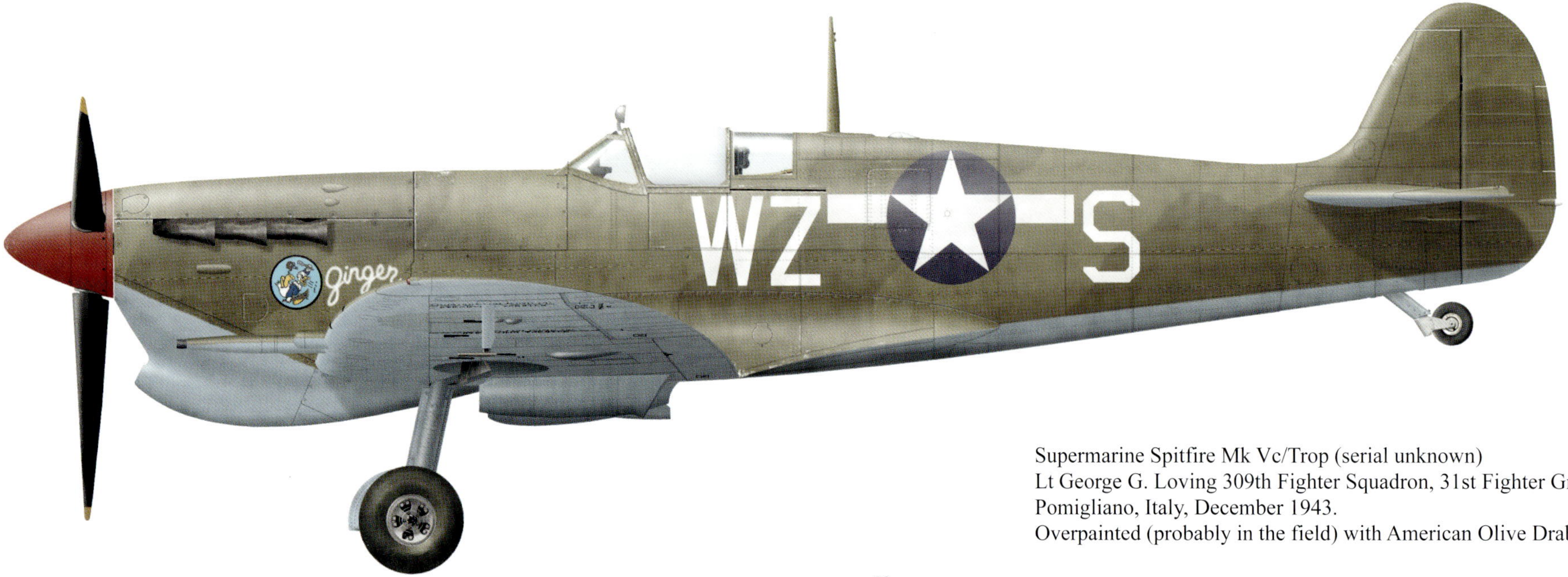

Supermarine Spitfire Mk Vc/Trop (serial unknown)
Lt George G. Loving 309th Fighter Squadron, 31st Fighter Group, USAAF
Pomigliano, Italy, December 1943.
Overpainted (probably in the field) with American Olive Drab colour.

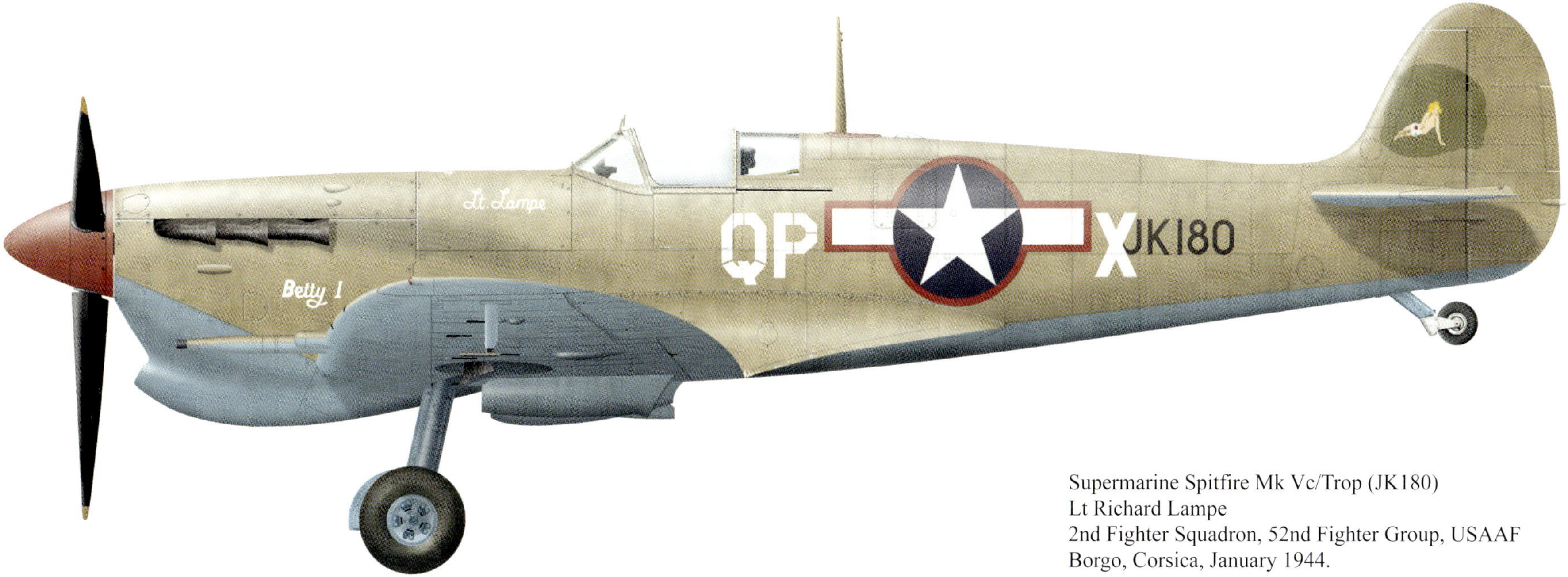

Supermarine Spitfire Mk Vc/Trop (JK180)
Lt Richard Lampe
2nd Fighter Squadron, 52nd Fighter Group, USAAF
Borgo, Corsica, January 1944.

Supermarine Spitfire Mk Vc/Trop (serial unknown)
Lt Richard Alexander
2nd Fighter Squadron, 52nd Fighter Group, USAAF
Borgo, Corsica, early 1944.

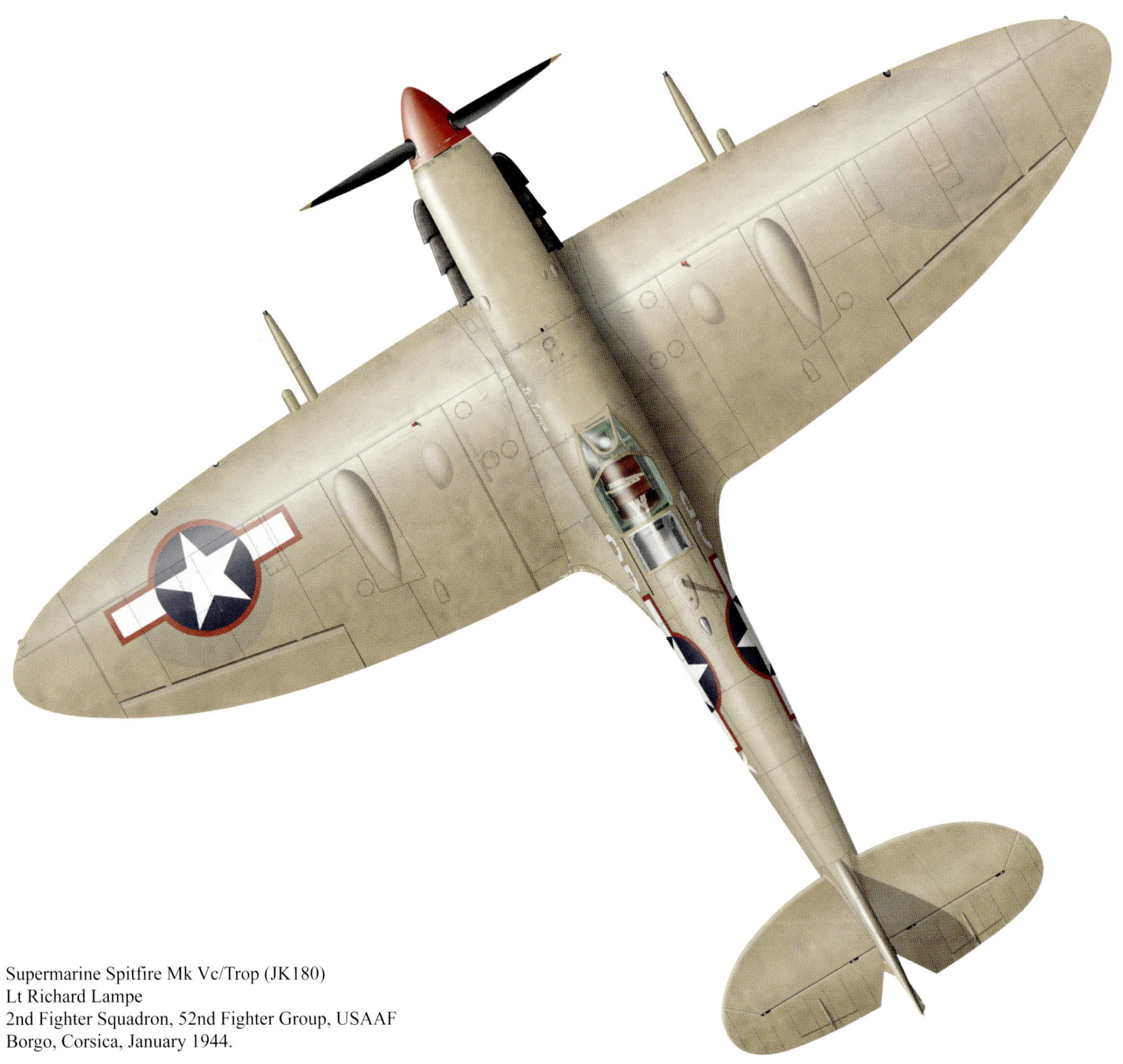

Supermarine Spitfire Mk Vc/Trop (JK180)
Lt Richard Lampe
2nd Fighter Squadron, 52nd Fighter Group, USAAF
Borgo, Corsica, January 1944.

Supermarine Spitfire Mk Vc/Trop (JK180)
Lt Richard Lampe
2nd Fighter Squadron, 52nd Fighter Group, USAAF
Borgo, Corsica, January 1944.

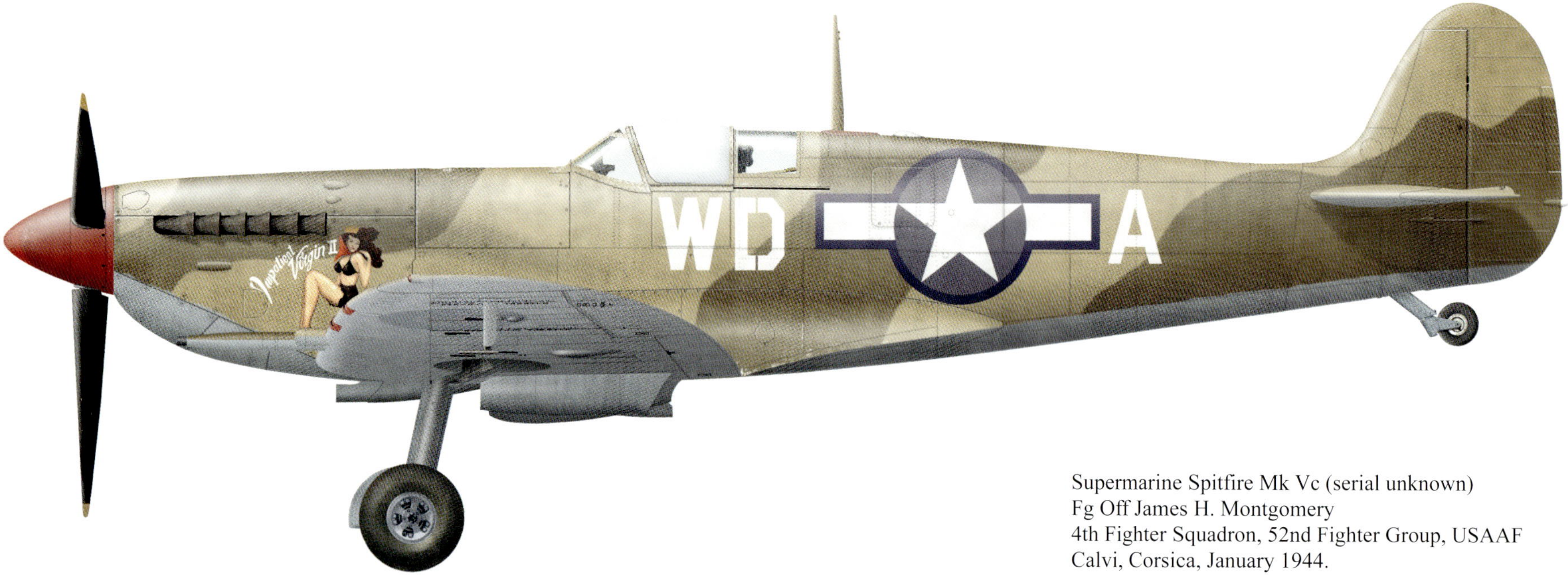

Supermarine Spitfire Mk Vc (serial unknown)
Fg Off James H. Montgomery
4th Fighter Squadron, 52nd Fighter Group, USAAF
Calvi, Corsica, January 1944.

Supermarine Spitfire Mk Vc (serial unknown)
Lt John Anderson, 5th Fighter Squadron, 52nd Fighter Group, USAAF
La Sabala, Algeria, February 1944.
Note non-standard three-colour camouflage.

Supermarine Spitfire Mk Vc/Trop (AR524)
Groupe de Chasse GC I/7, Armée de l'Air
Djidjeli-Taher, Algeria, early 1944.

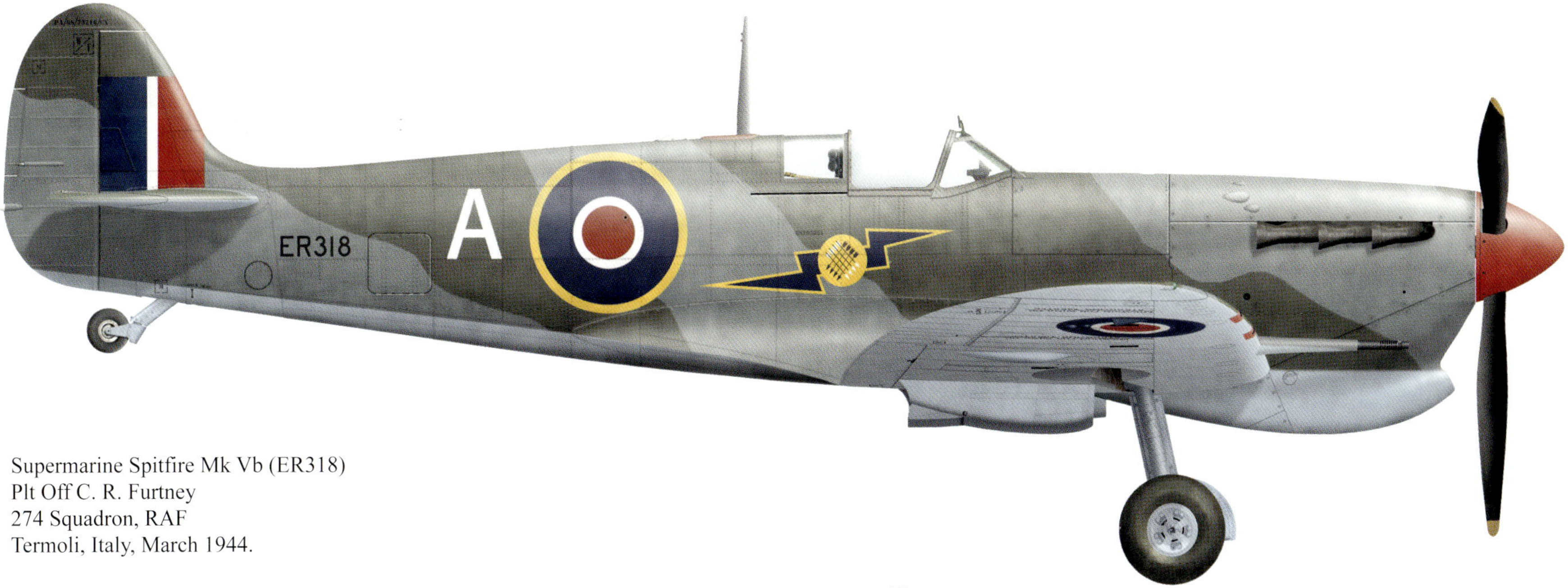

Supermarine Spitfire Mk Vb (ER318)
Plt Off C. R. Furtney
274 Squadron, RAF
Termoli, Italy, March 1944.

Supermarine Spitfire Mk Vb (BK729)
VCS-7 Squadron, US Navy, Lee-on-Solent, UK, June 1944.
Although this aircraft was used by a US Navy unit to spot for the fleet on D-Day it still retained British national insignia.

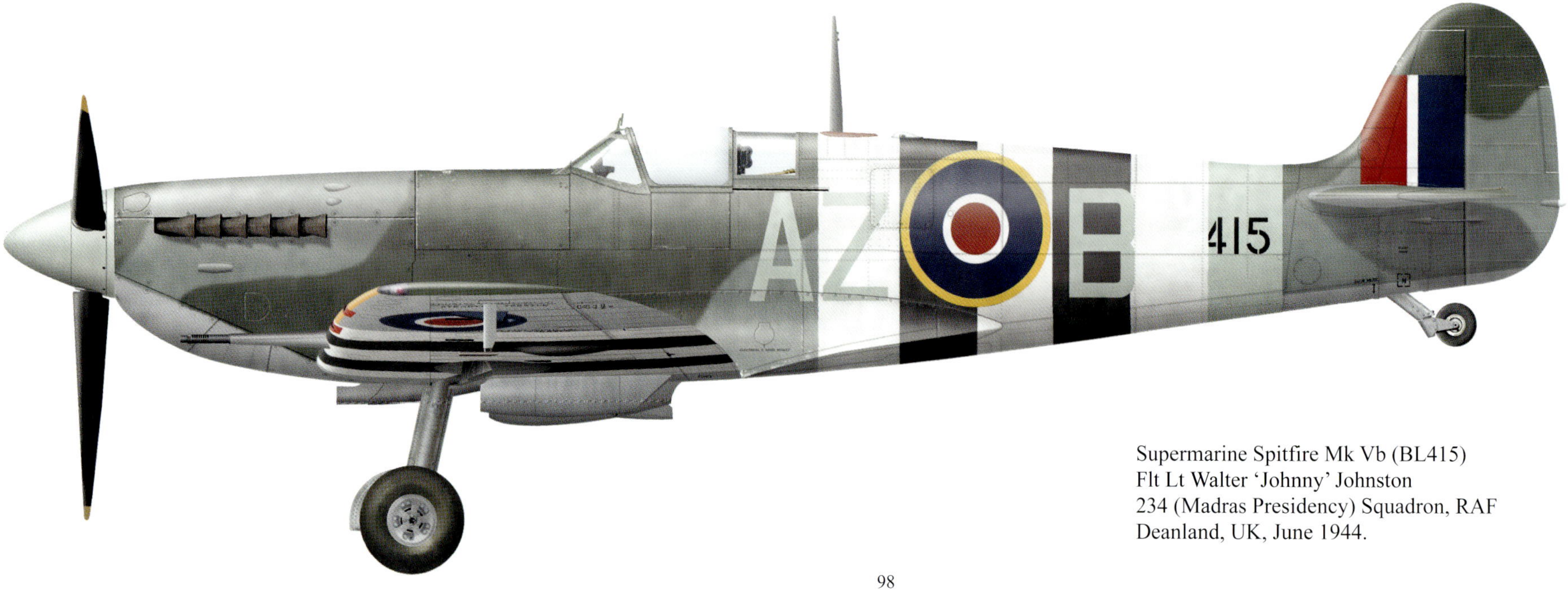

Supermarine Spitfire Mk Vb (BL415)
Flt Lt Walter 'Johnny' Johnston
234 (Madras Presidency) Squadron, RAF
Deanland, UK, June 1944.

Supermarine Spitfire LF Mk Vc (AB509)
Wg Cdr John M. Checketts
142 Wing, RAF
Horne, UK, June 1944.

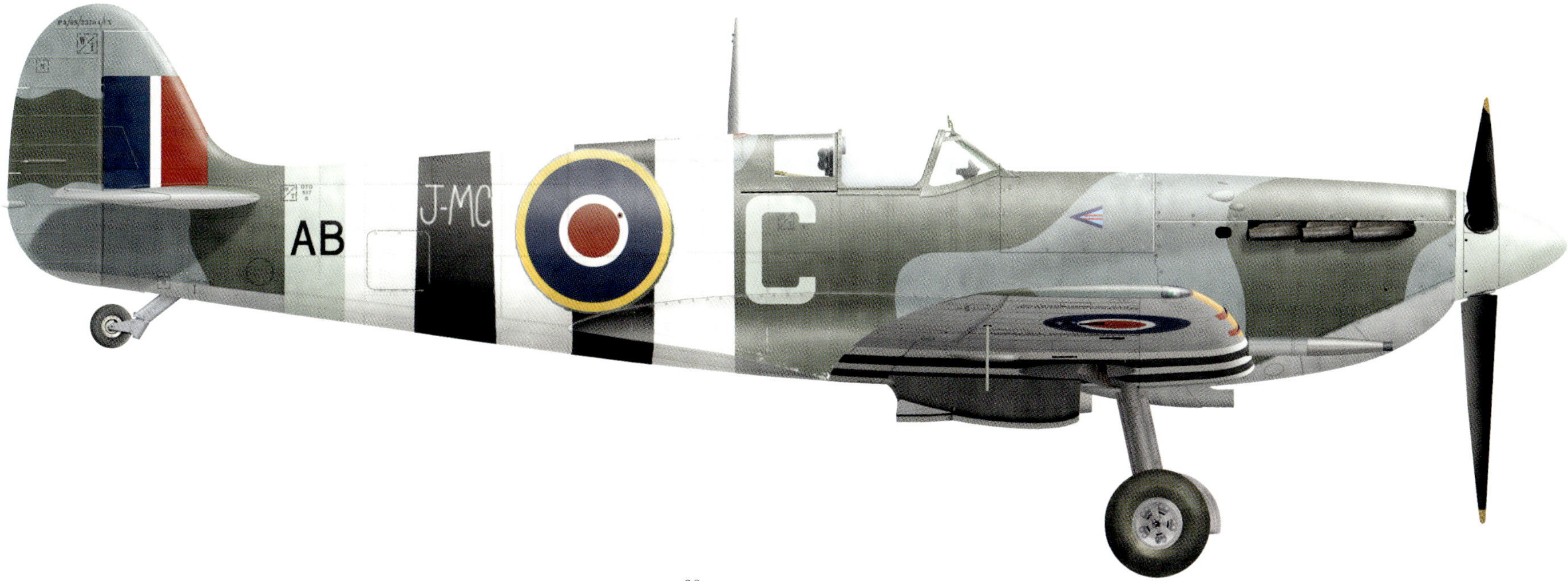

Supermarine Spitfire LF Mk Vc (AB509)
Wg Cdr John M. Checketts
142 Wing, RAF
Horne, UK, June 1944.

Supermarine Spitfire LF Mk Vc (AB509)
Wg Cdr John M. Checketts
142 Wing, RAF
Horne, UK, June 1944.

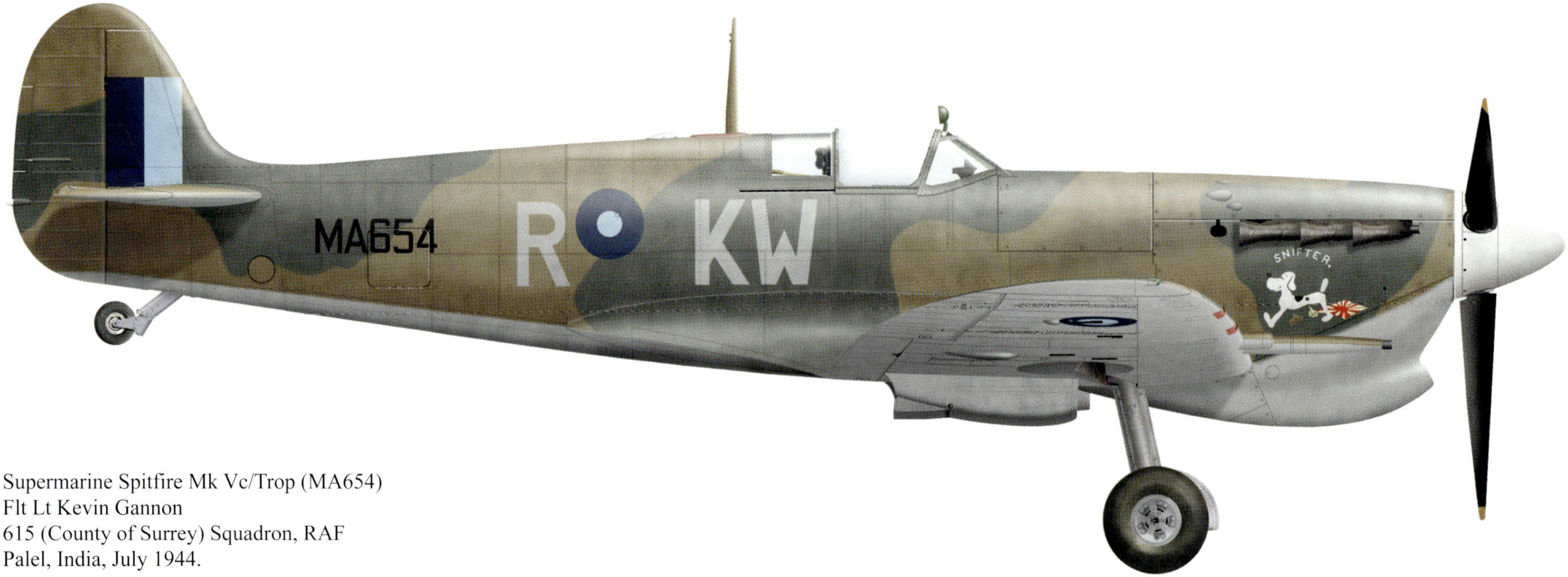

Supermarine Spitfire Mk Vc/Trop (MA654)
Flt Lt Kevin Gannon
615 (County of Surrey) Squadron, RAF
Palel, India, July 1944.

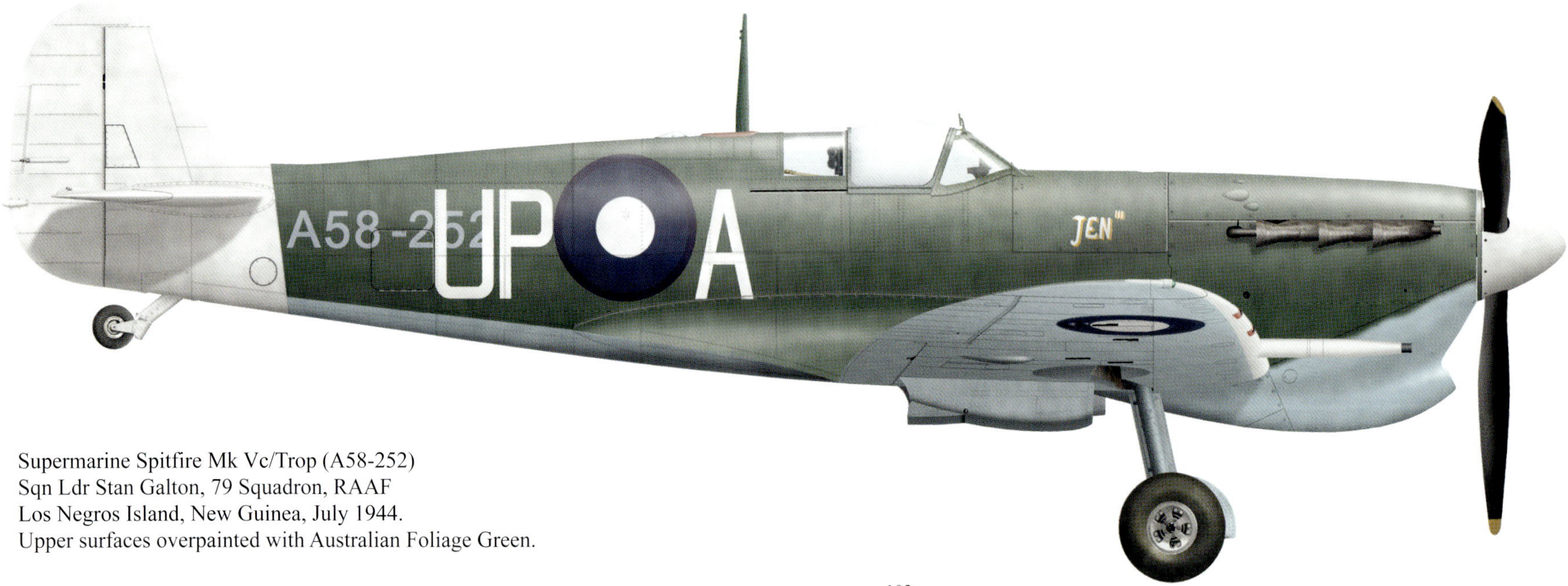

Supermarine Spitfire Mk Vc/Trop (A58-252)
Sqn Ldr Stan Galton, 79 Squadron, RAAF
Los Negros Island, New Guinea, July 1944.
Upper surfaces overpainted with Australian Foliage Green.

Supermarine Spitfire Mk Vc/Trop (JK891)
335 (Greek) Squadron, RAF
Hassani, Greece, February 1945.

Supermarine Spitfire Mk Vc/Trop (A58-250)
85 Squadron, RAAF
Pearce, Australia, May 1945.

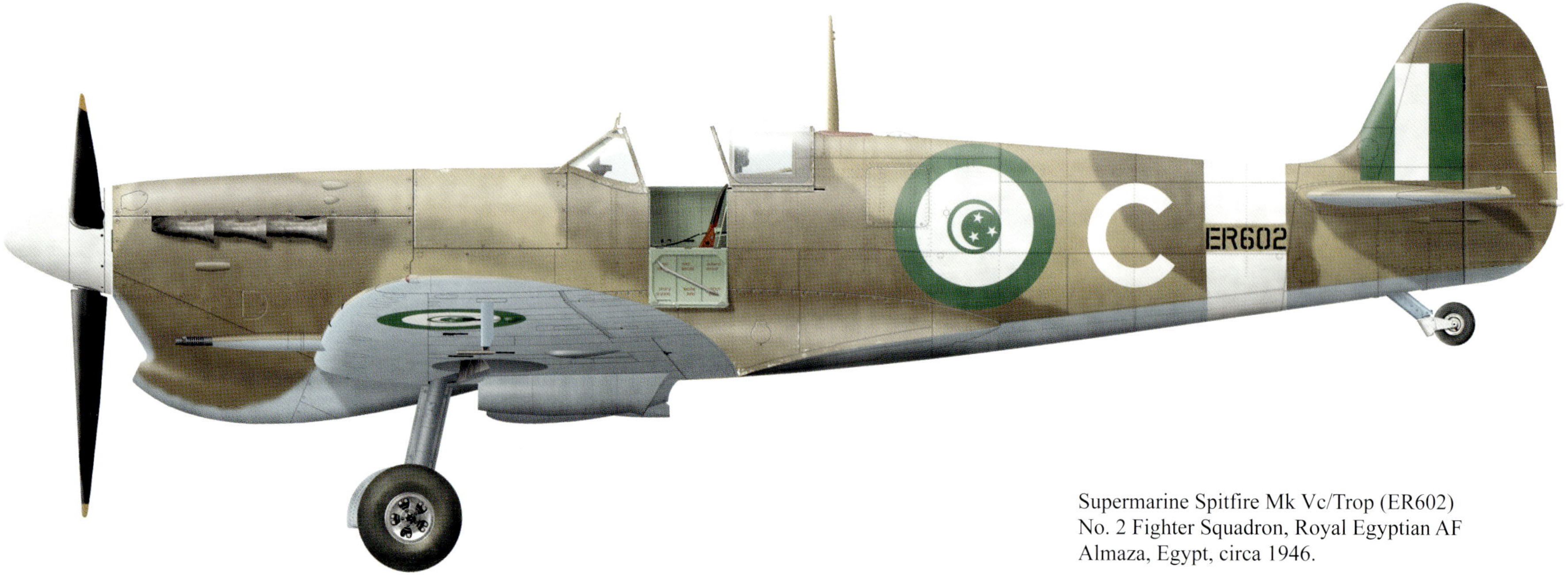

Supermarine Spitfire Mk Vc/Trop (ER602)
No. 2 Fighter Squadron, Royal Egyptian AF
Almaza, Egypt, circa 1946.

Supermarine Spitfire Mk Vb (serial unknown)
2. Esquadrilha de Caça, Aviação Militar Portuguesa
Ota, Portugal, October 1950.

Spitfire Mk VI

Supermarine Spitfire Mk VI (BR326)
Plt Off Jean Maridor
91 (Nigeria) Squadron, RAF
Hawkinge, UK, August 1942.

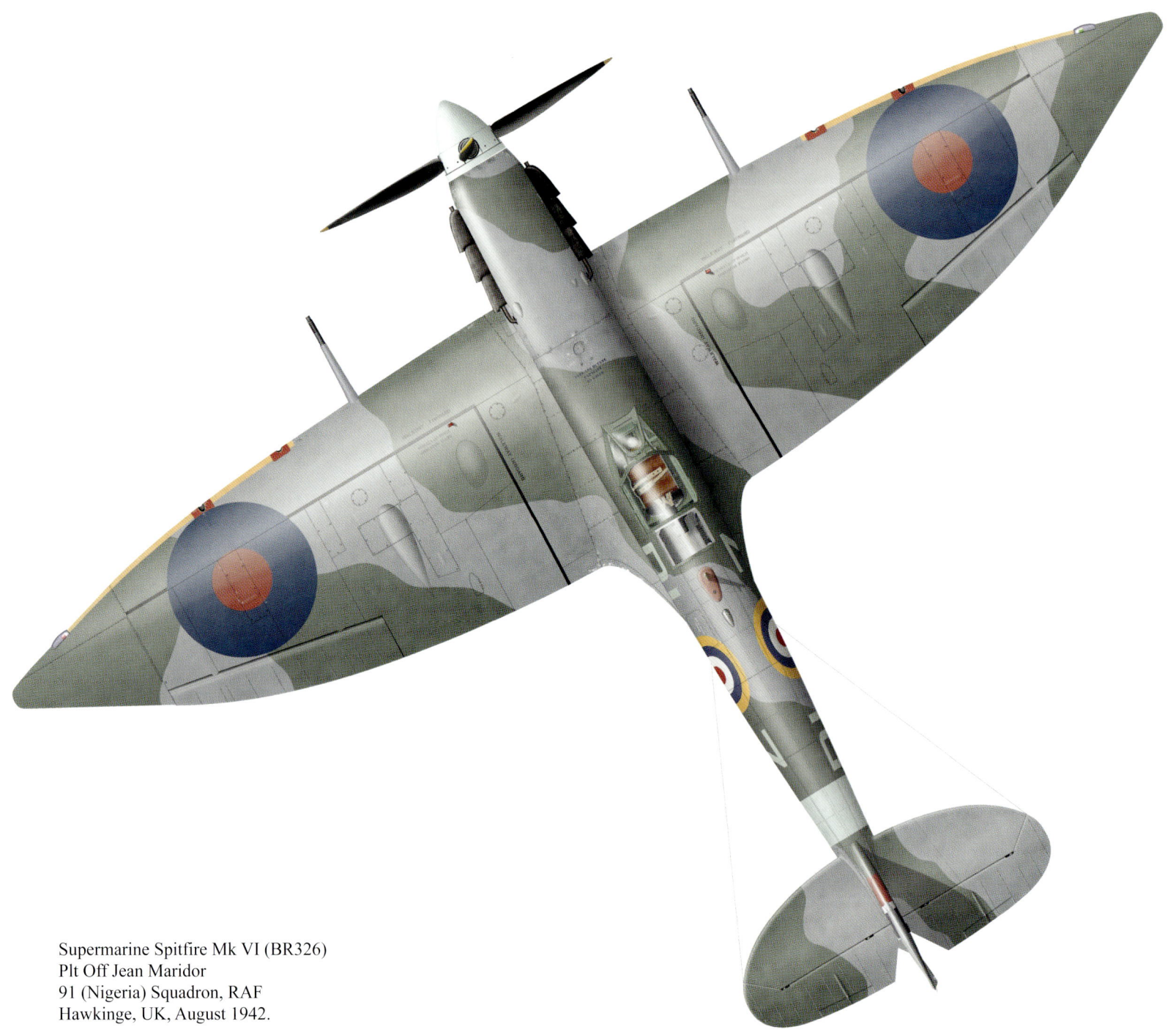

Supermarine Spitfire Mk VI (BR326)
Plt Off Jean Maridor
91 (Nigeria) Squadron, RAF
Hawkinge, UK, August 1942.

Supermarine Spitfire Mk VI (BR326)
Plt Off Jean Maridor
91 (Nigeria) Squadron, RAF
Hawkinge, UK, August 1942.

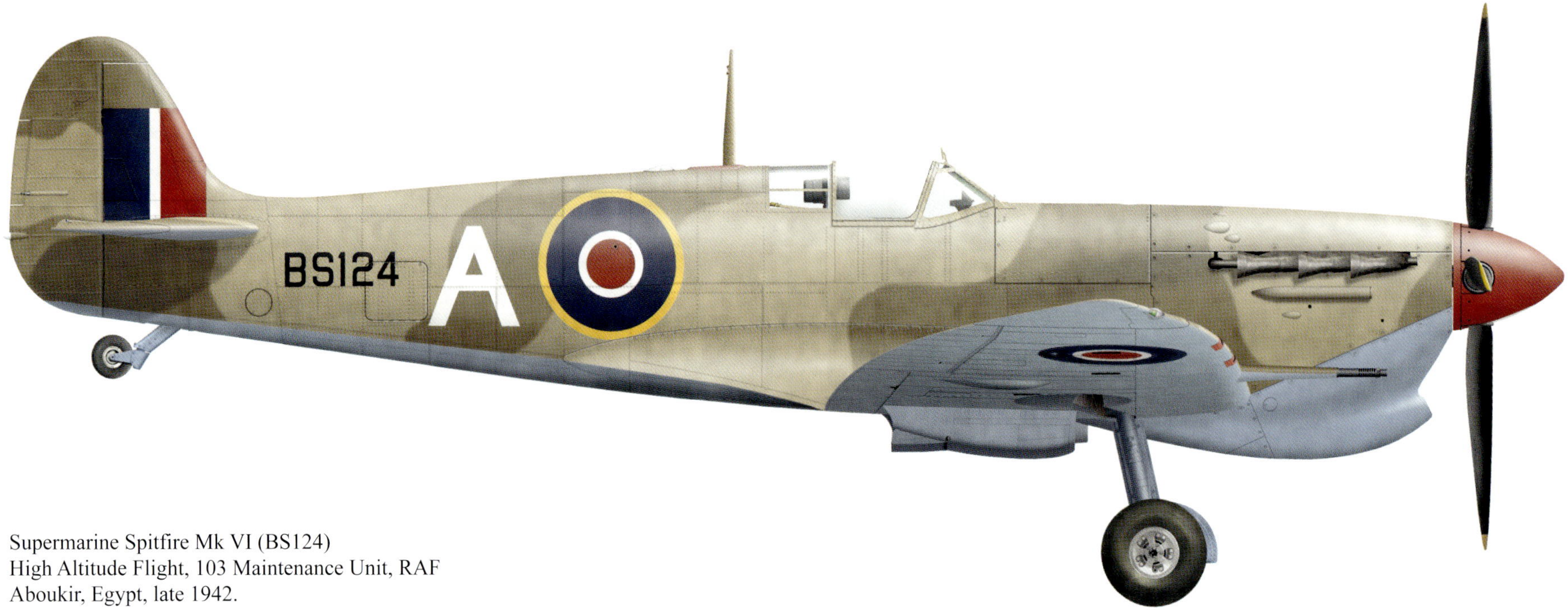

Supermarine Spitfire Mk VI (BS124)
High Altitude Flight, 103 Maintenance Unit, RAF
Aboukir, Egypt, late 1942.

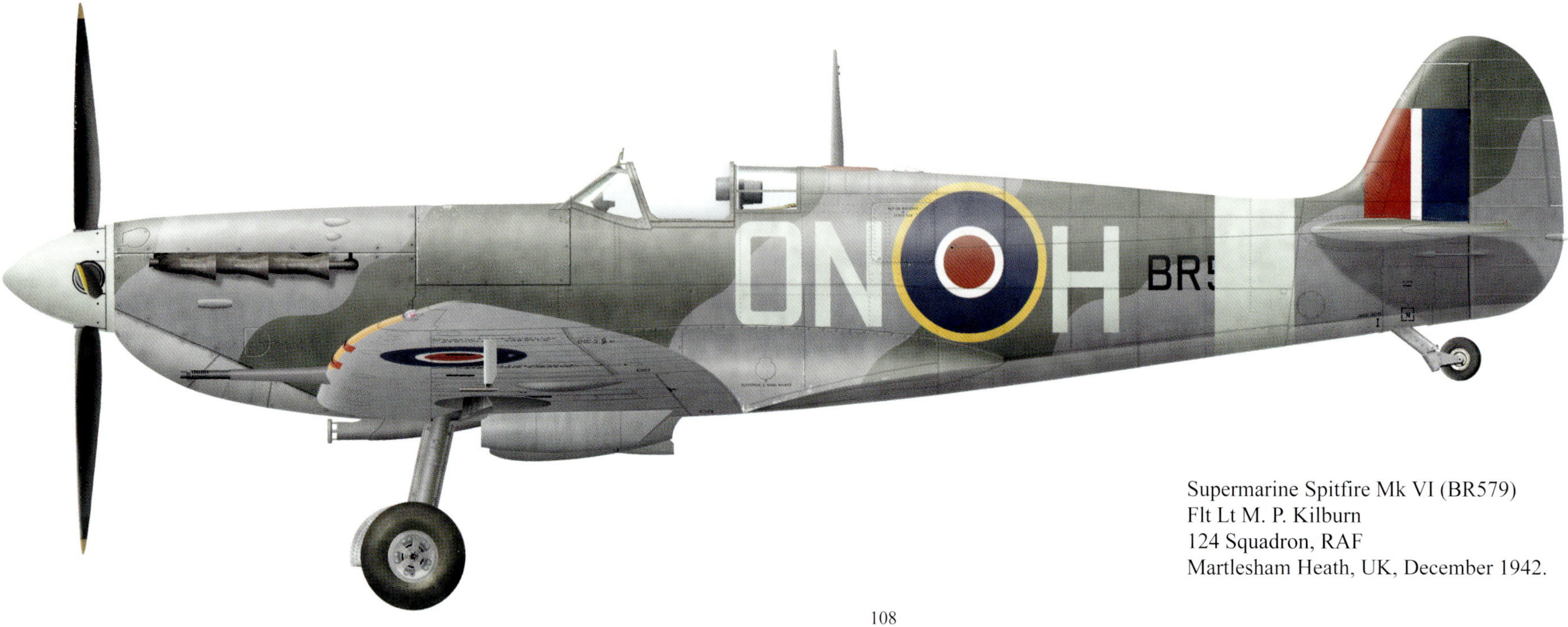

Supermarine Spitfire Mk VI (BR579)
Flt Lt M. P. Kilburn
124 Squadron, RAF
Martlesham Heath, UK, December 1942.

Spitfire Mk VII

Supermarine Spitfire HF Mk VIII (EN285)
Flt Sgt Desmond Kelly
124 (Baroda) Squadron, RAF
West Malling, UK, summer 1943.

Supermarine Spitfire HF Mk VII (MD114)
453 Squadron, RAAF
Skeabrae, UK, January 1944.

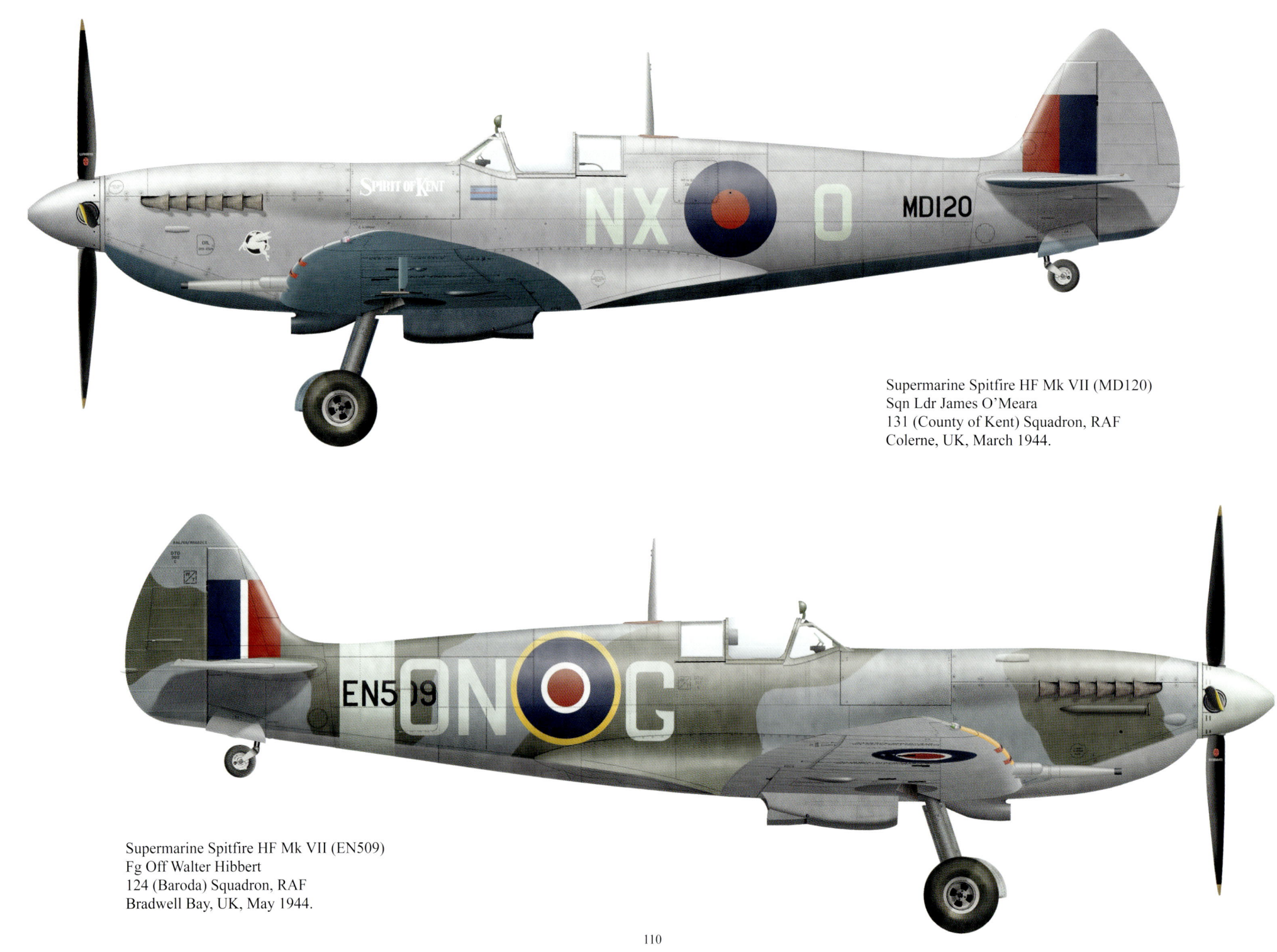

Supermarine Spitfire HF Mk VII (MD120)
Sqn Ldr James O'Meara
131 (County of Kent) Squadron, RAF
Colerne, UK, March 1944.

Supermarine Spitfire HF Mk VII (EN509)
Fg Off Walter Hibbert
124 (Baroda) Squadron, RAF
Bradwell Bay, UK, May 1944.

Supermarine Spitfire HF Mk VII
Wg Cdr Peter Brothers
Culmhead Wing, RAF
Culmhead, UK, June 1944.

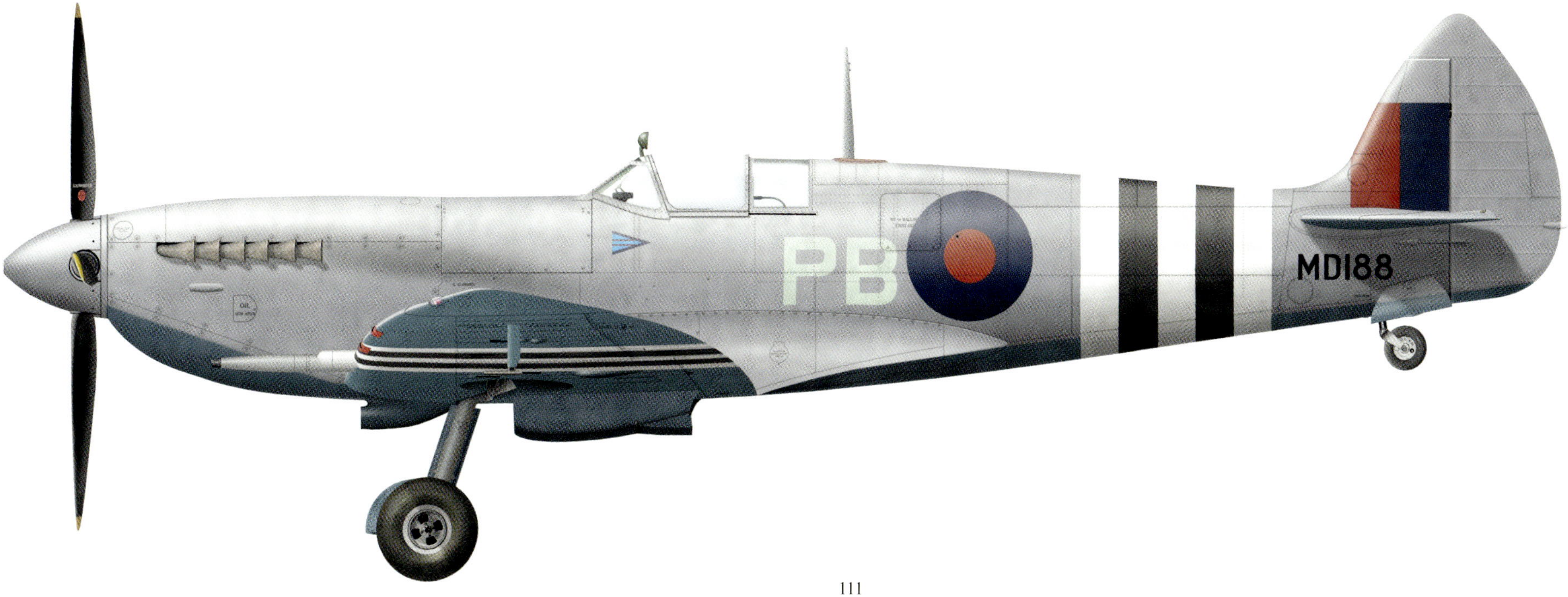

Supermarine Spitfire HF Mk VII
Wg Cdr Peter Brothers
Culmhead Wing, RAF
Culmhead, UK, June 1944.

Supermarine Spitfire HF Mk VII
Wg Cdr Peter Brothers
Culmhead Wing, RAF
Culmhead, UK, June 1944.

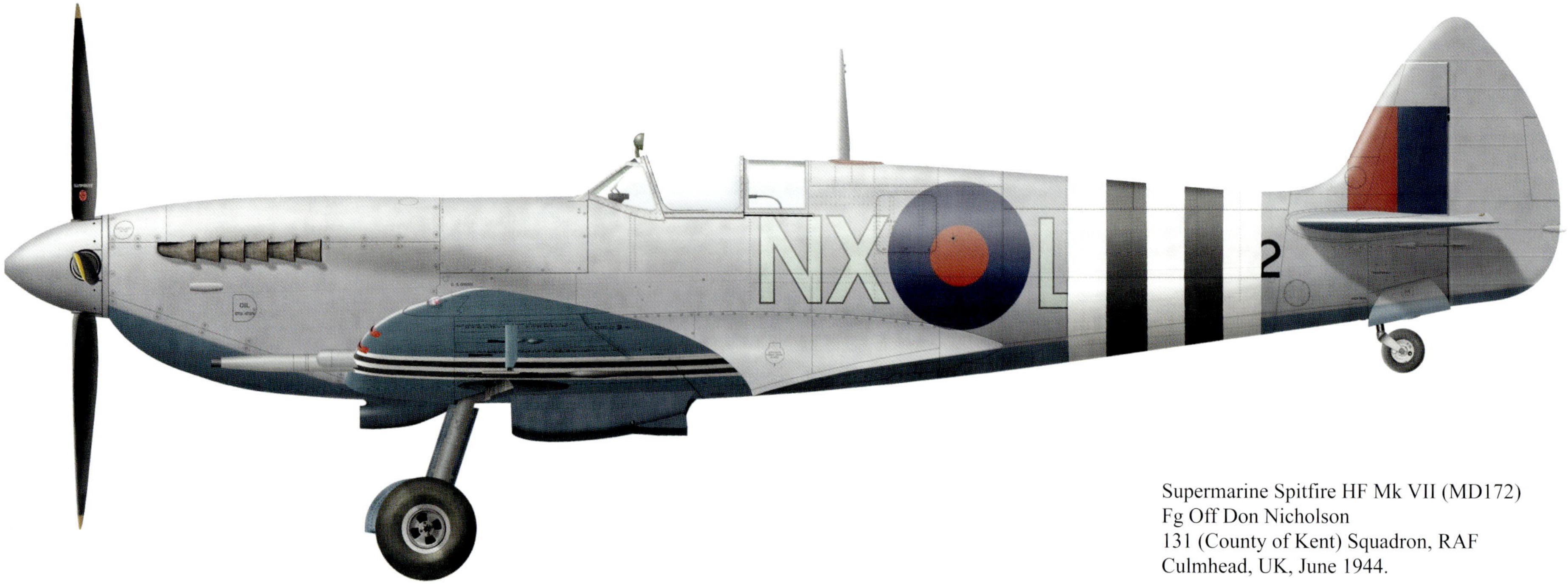

Supermarine Spitfire HF Mk VII (MD172)
Fg Off Don Nicholson
131 (County of Kent) Squadron, RAF
Culmhead, UK, June 1944.

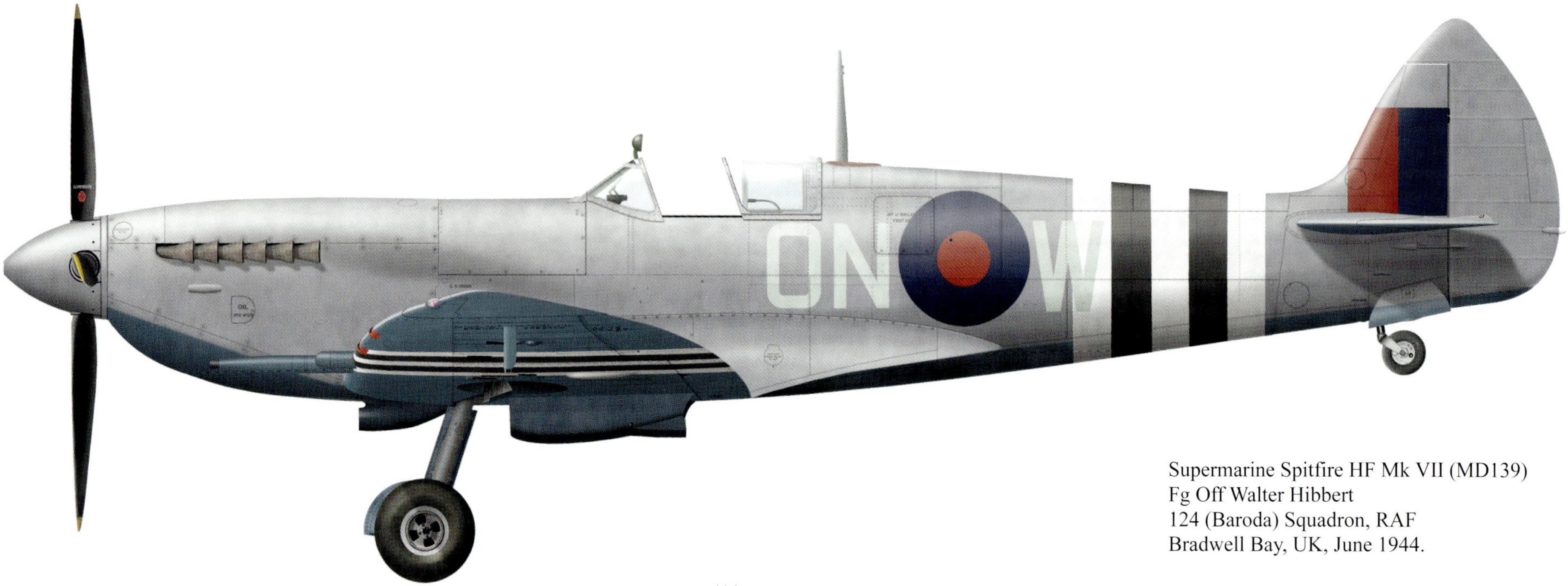

Supermarine Spitfire HF Mk VII (MD139)
Fg Off Walter Hibbert
124 (Baroda) Squadron, RAF
Bradwell Bay, UK, June 1944.

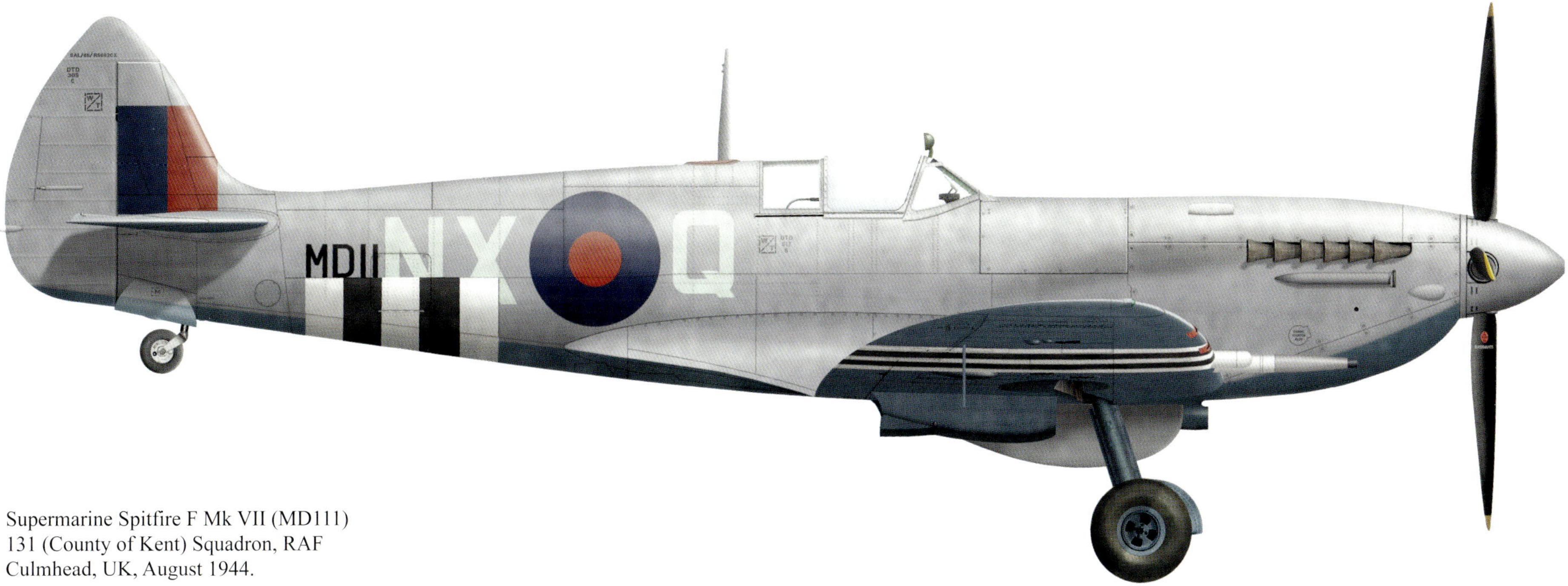

Supermarine Spitfire F Mk VII (MD111)
131 (County of Kent) Squadron, RAF
Culmhead, UK, August 1944.

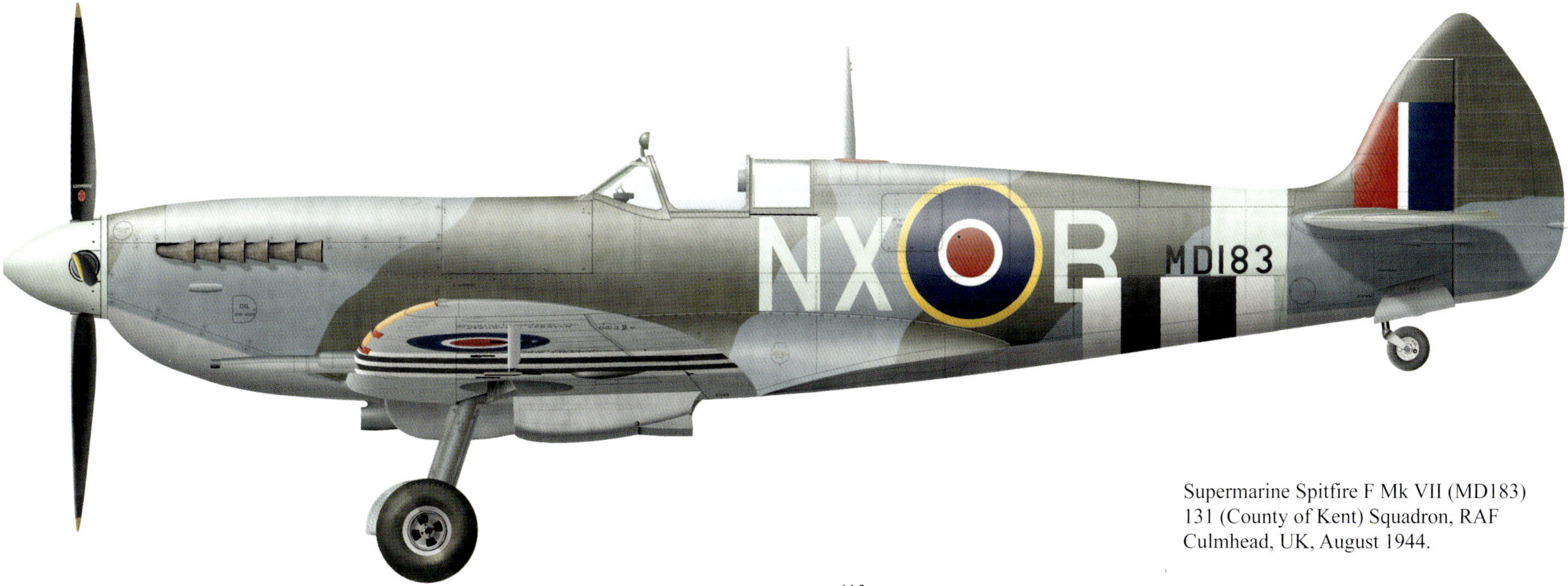

Supermarine Spitfire F Mk VII (MD183)
131 (County of Kent) Squadron, RAF
Culmhead, UK, August 1944.

Supermarine Spitfire HF Mk VII (MD182)
Flt Lt Jack Cleland
616 (South Yorkshire) Squadron, RAF
Manston, UK, September 1944.

Spitfire Mk VIII

Supermarine Spitfire HF Mk VIII (JF447)
Sqn Ldr Stanislaw Skalski
601 (County of London) Squadron, RAF
Lentini, Sicily, August 1943.

Supermarine Spitfire F Mk VIII (JF469)
Flt Lt Albert Houle
417 Squadron, RCAF
Gioia del Colle, Italy, October 1943.

Supermarine Spitfire HF Mk VIII (JF421)
1 Squadron, SAAF
Palata, Italy, November 1943.

Supermarine Spitfire F MK VIII (serial unknown)
Col C. M. McCorkle
31st Fighter Group, USAAF
Castel Volturno, Italy, January 1944.

Supermarine Spitfire HF Mk VIII (JF476)
Lt John 'Johnny' Gasson
92 (East India) Squadron, RAF
Marcianise, Italy, January 1944.

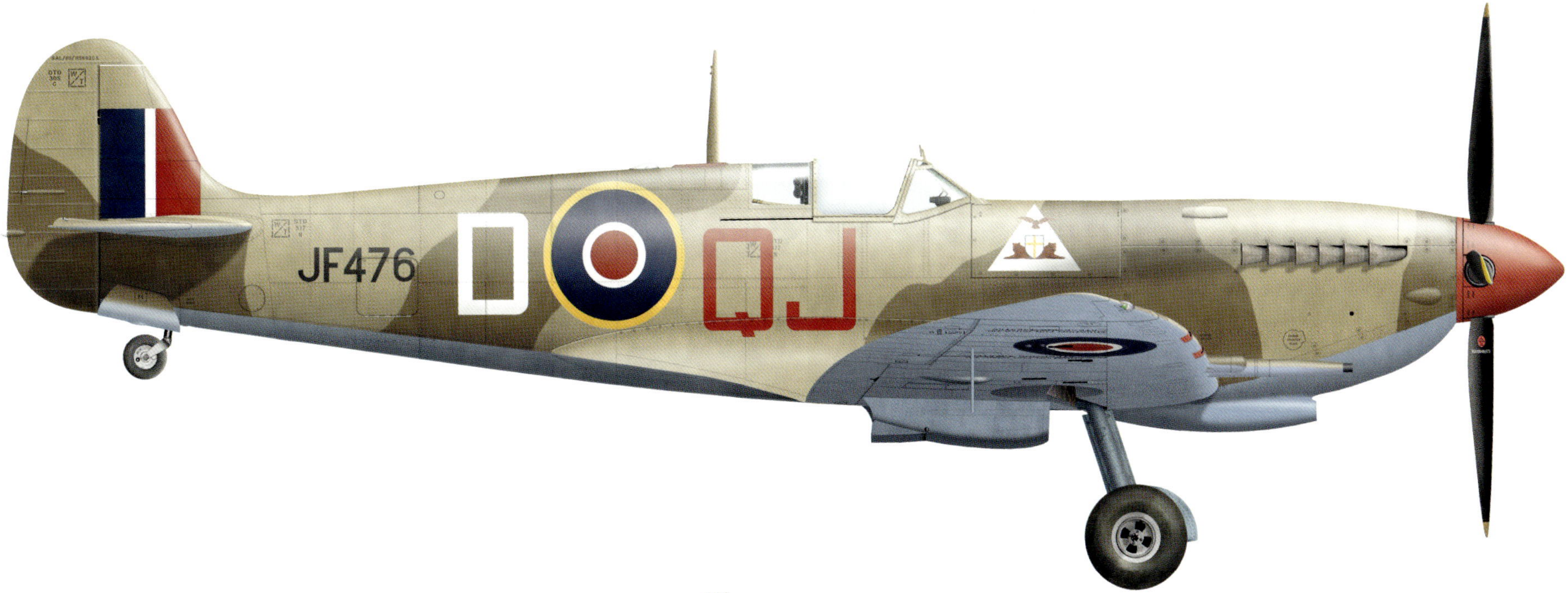

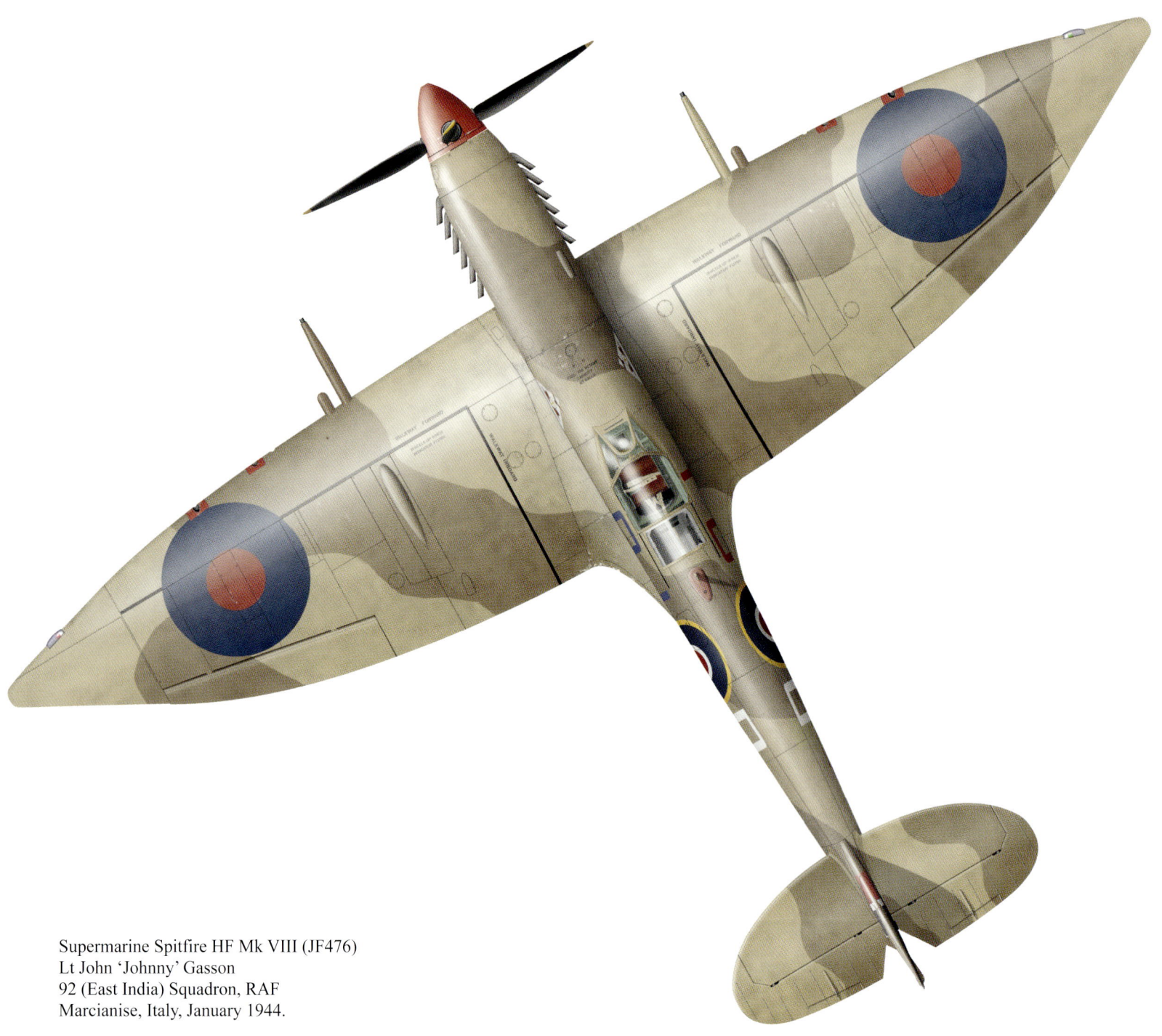

Supermarine Spitfire HF Mk VIII (JF476)
Lt John 'Johnny' Gasson
92 (East India) Squadron, RAF
Marcianise, Italy, January 1944.

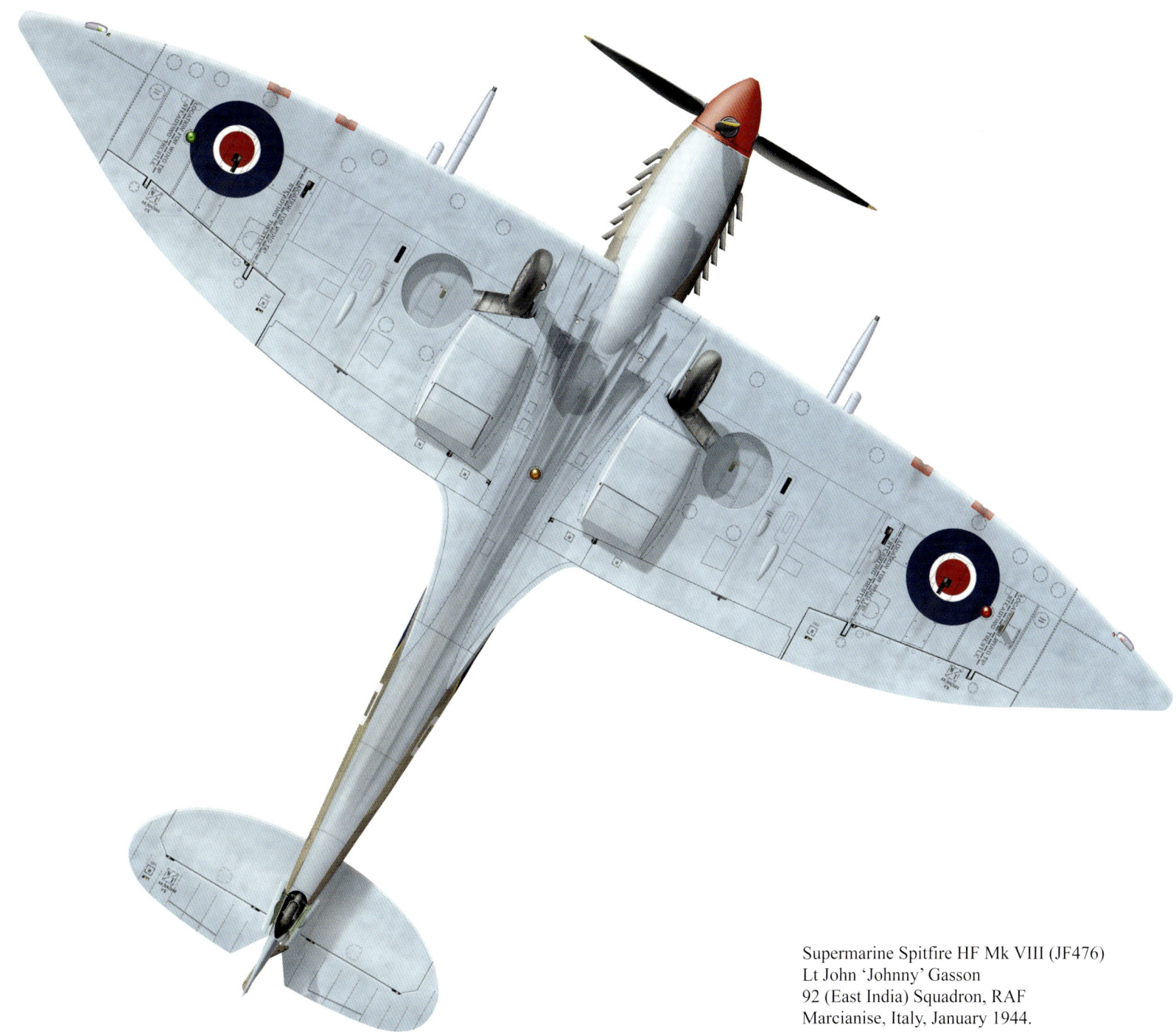

Supermarine Spitfire HF Mk VIII (JF476)
Lt John 'Johnny' Gasson
92 (East India) Squadron, RAF
Marcianise, Italy, January 1944.

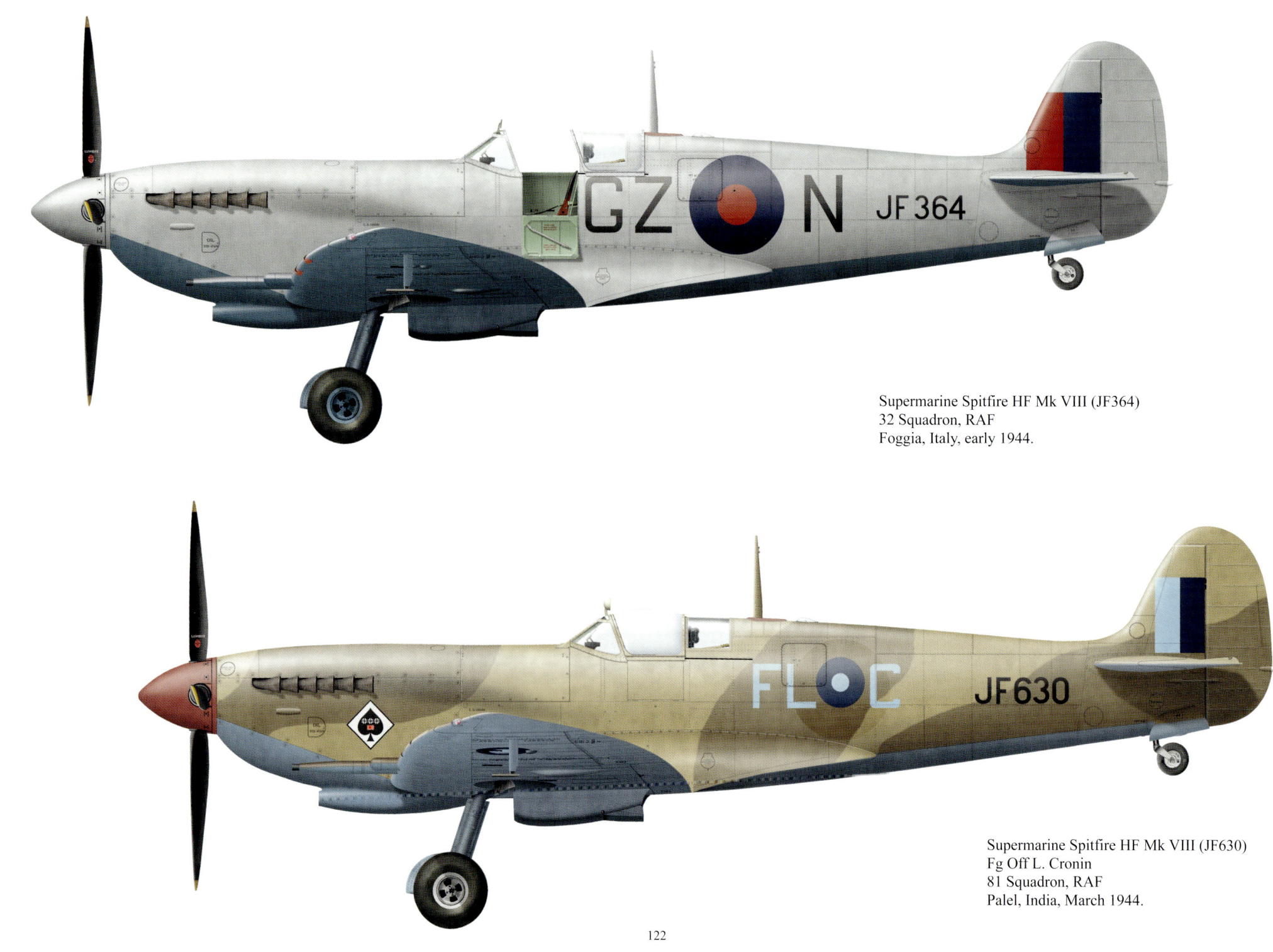

Supermarine Spitfire HF Mk VIII (JF364)
32 Squadron, RAF
Foggia, Italy, early 1944.

Supermarine Spitfire HF Mk VIII (JF630)
Fg Off L. Cronin
81 Squadron, RAF
Palel, India, March 1944.

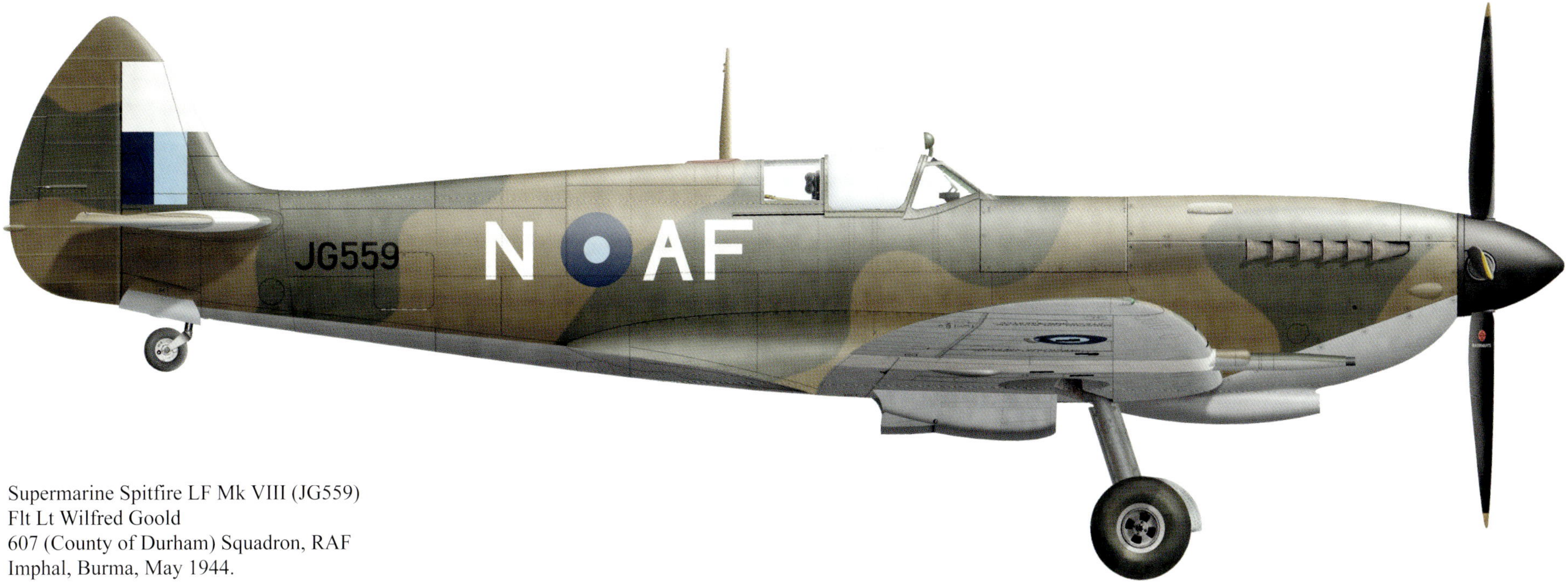

Supermarine Spitfire LF Mk VIII (JG559)
Flt Lt Wilfred Goold
607 (County of Durham) Squadron, RAF
Imphal, Burma, May 1944.

Supermarine Spitfire LF Mk VIII (A58-317)
Sgt J Blair
54 Squadron, RAF
Livingstone, Australia, summer 1944.

Supermarine Spitfire LF Mk VIII (MT775)
Sqn Ldr Neville Duke
145 Squadron, RAF
Loreto, Italy, July 1944.

Supermarine Spitfire LF Mk VIII (MT714)
Flt Lt A. W. Guest
43 (Fighting Cocks) Squadron, RAF
Ramatuelle, France, August 1944.

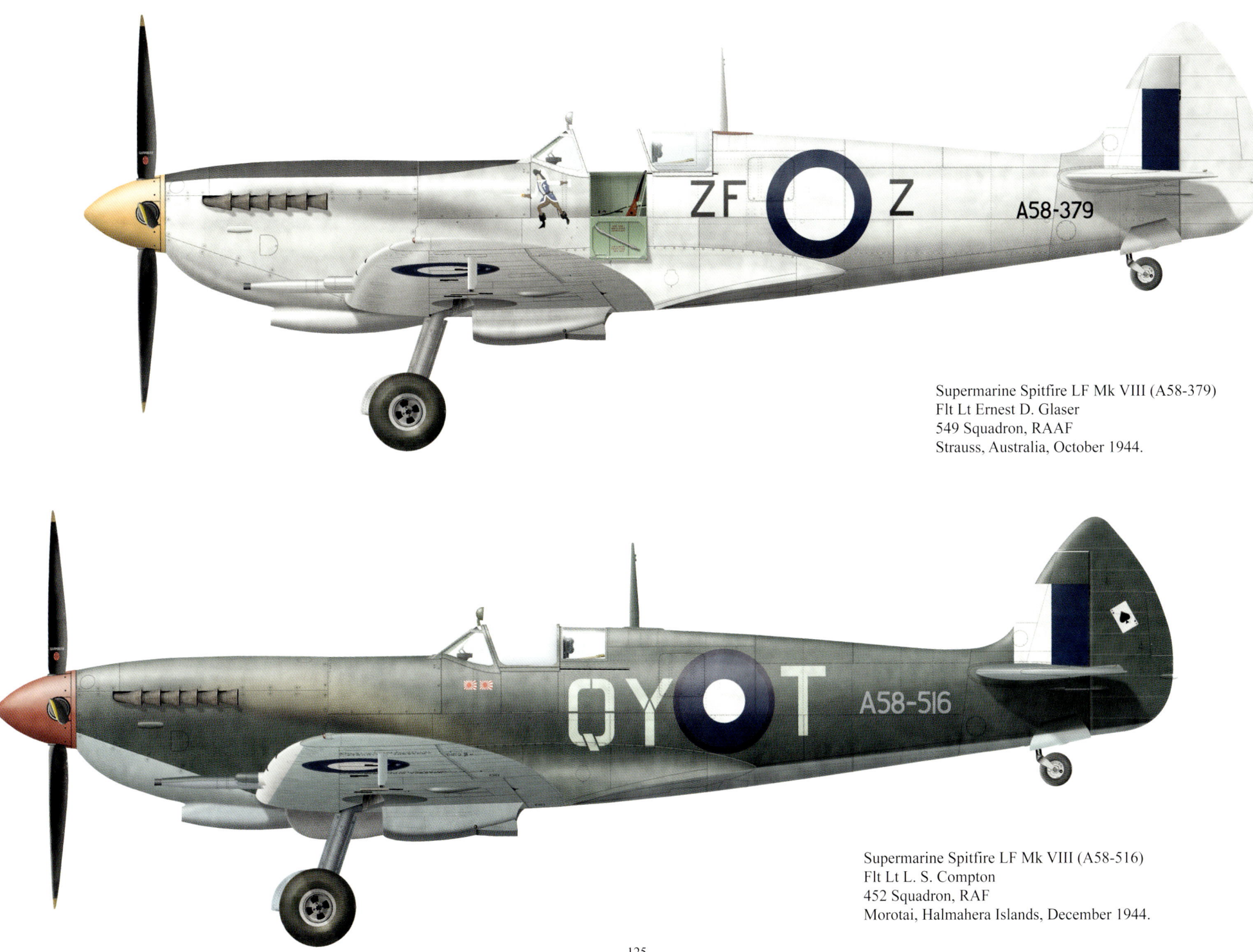

Supermarine Spitfire LF Mk VIII (A58-379)
Flt Lt Ernest D. Glaser
549 Squadron, RAAF
Strauss, Australia, October 1944.

Supermarine Spitfire LF Mk VIII (A58-516)
Flt Lt L. S. Compton
452 Squadron, RAF
Morotai, Halmahera Islands, December 1944.

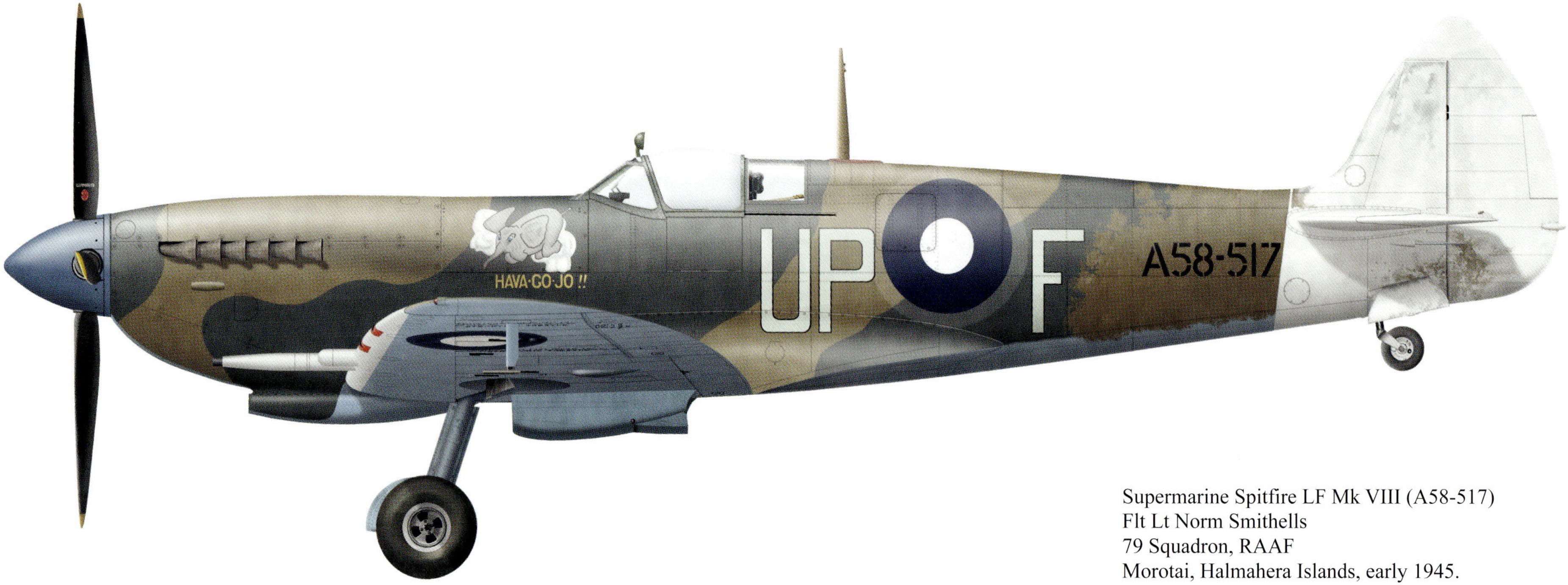

Supermarine Spitfire LF Mk VIII (A58-517)
Flt Lt Norm Smithells
79 Squadron, RAAF
Morotai, Halmahera Islands, early 1945.

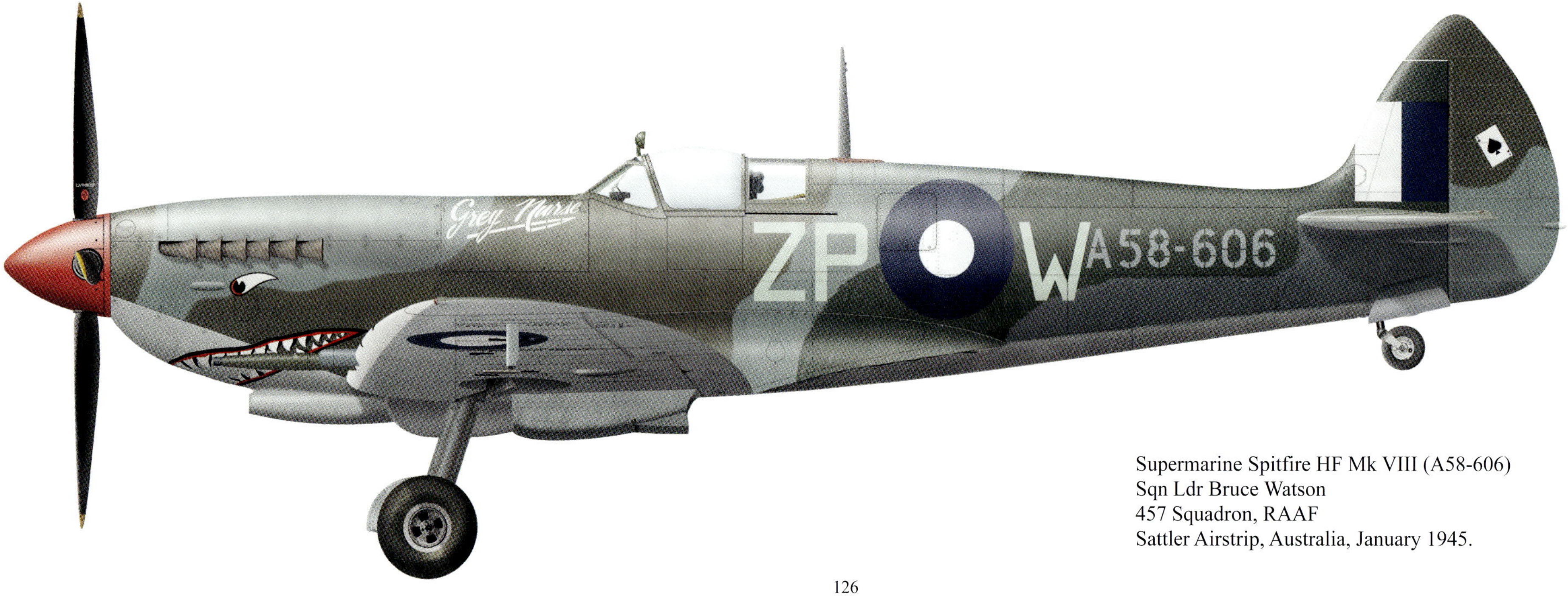

Supermarine Spitfire HF Mk VIII (A58-606)
Sqn Ldr Bruce Watson
457 Squadron, RAAF
Sattler Airstrip, Australia, January 1945.

Supermarine Spitfire LF Mk VIII (MT507)
Fg Off Len A. Smith
152 (Hyderabad) Squadron, RAF
Sinthe, Burma, March 1945.

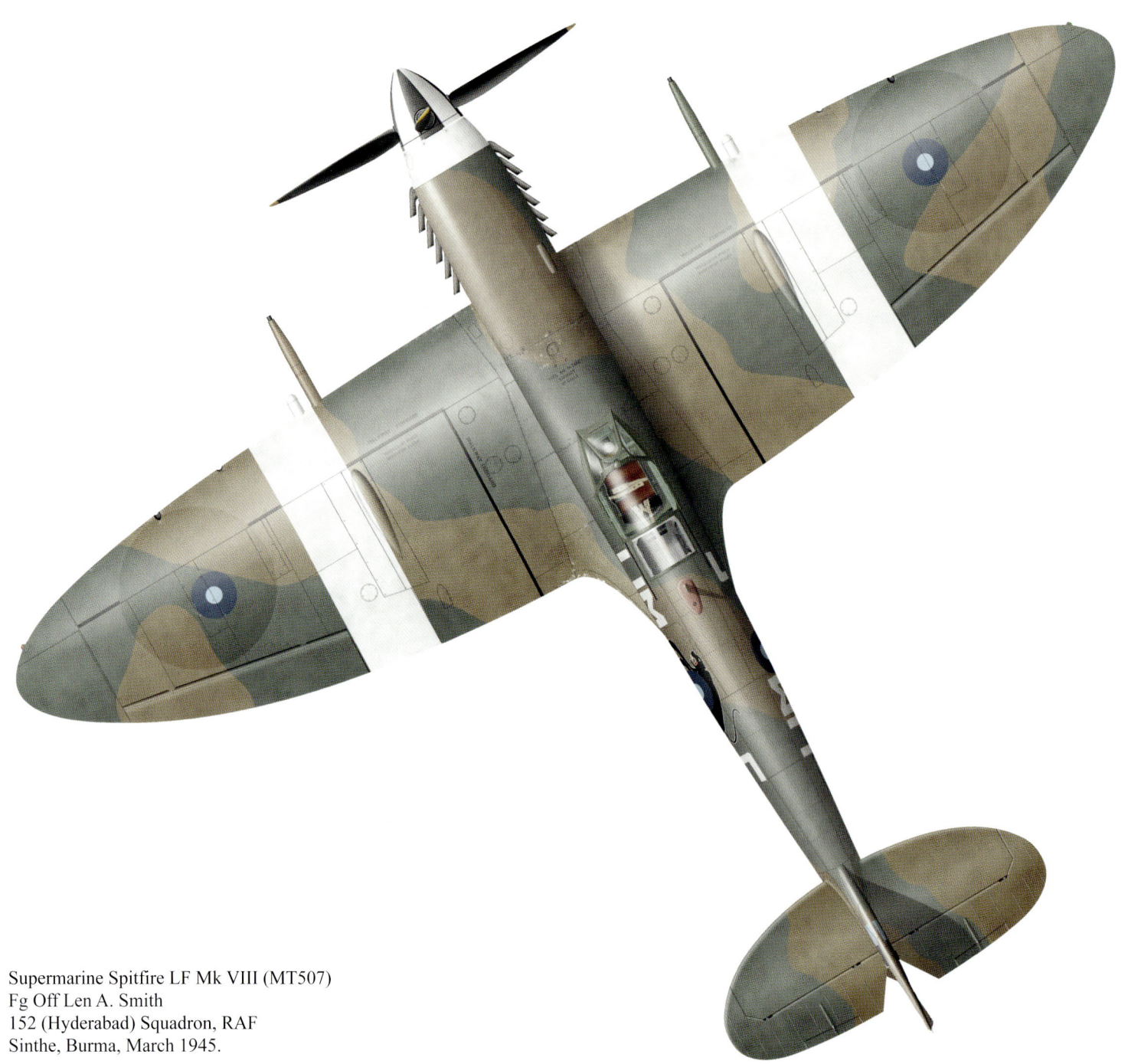

Supermarine Spitfire LF Mk VIII (MT507)
Fg Off Len A. Smith
152 (Hyderabad) Squadron, RAF
Sinthe, Burma, March 1945.

Supermarine Spitfire LF Mk VIII (MT507)
Fg Off Len A. Smith
152 (Hyderabad) Squadron, RAF
Sinthe, Burma, March 1945.

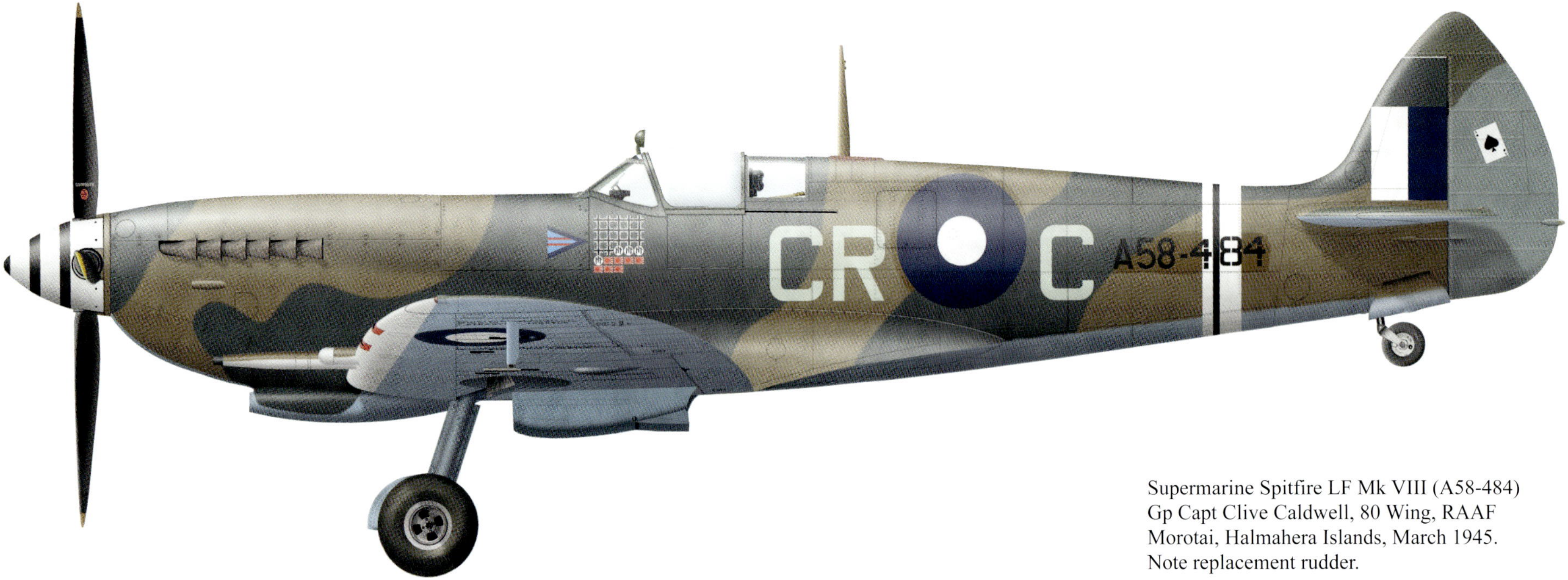

Supermarine Spitfire LF Mk VIII (A58-484)
Gp Capt Clive Caldwell, 80 Wing, RAAF
Morotai, Halmahera Islands, March 1945.
Note replacement rudder.

Supermarine Spitfire HF Mk VIII (A58-602)
Wg Cdr Bobby Gibbs
80 Wing, RAAF
Morotai, Halmahera Island, April 1945.

Supermarine Spitfire LF Mk VIII (A58-543)
Wt Off Hubert Eccleston
79 Squadron, RAAF
Morotai, Halmahera Islands, May 1945.

Supermarine Spitfire LF Mk VIII (A58-543)
Wt Off Hubert Eccleston
79 Squadron, RAAF
Morotai, Halmahera Islands, May 1945.

Supermarine Spitfire LF Mk VIII (A58-543)
Wt Off Hubert Eccleston
79 Squadron, RAAF
Morotai, Halmahera Islands, May 1945.

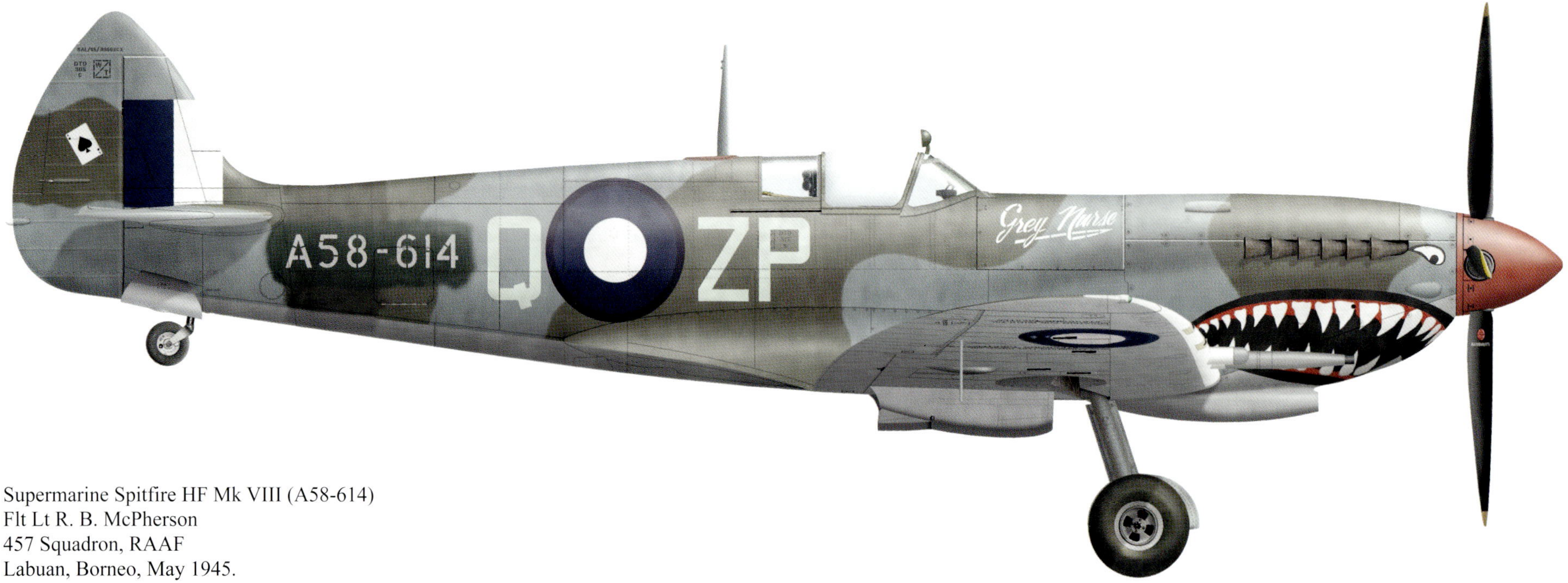

Supermarine Spitfire HF Mk VIII (A58-614)
Flt Lt R. B. McPherson
457 Squadron, RAAF
Labuan, Borneo, May 1945.

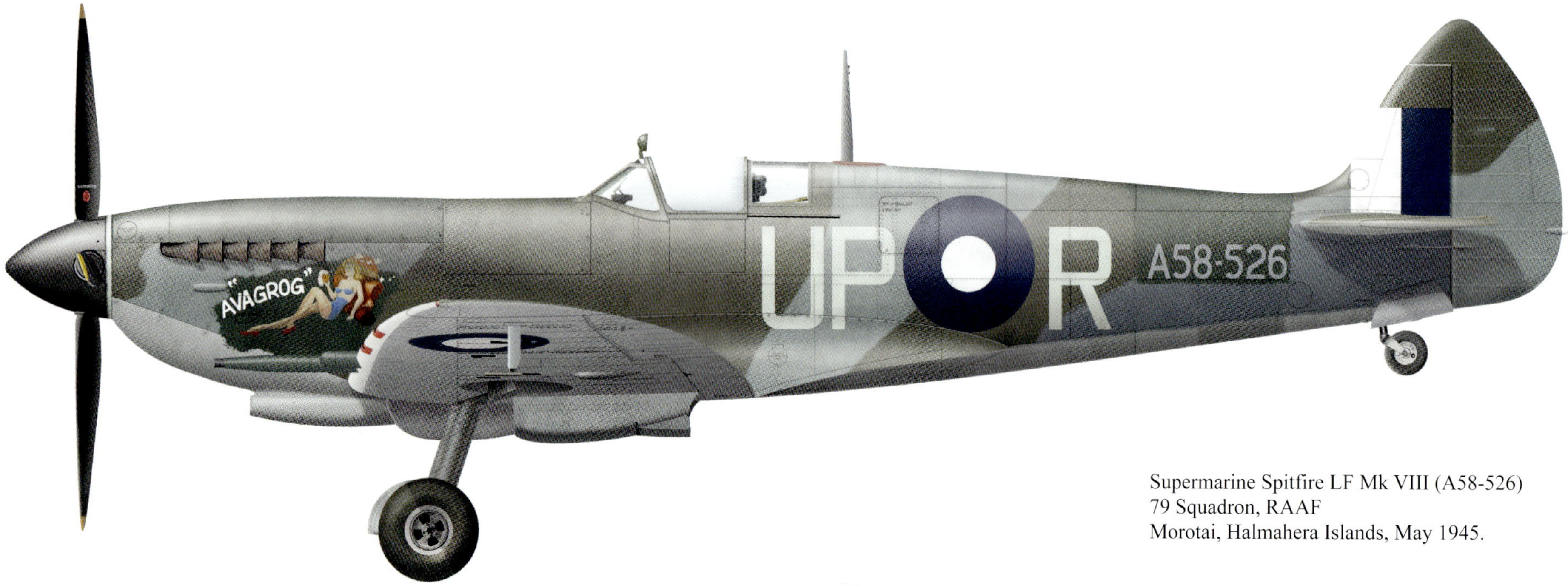

Supermarine Spitfire LF Mk VIII (A58-526)
79 Squadron, RAAF
Morotai, Halmahera Islands, May 1945.

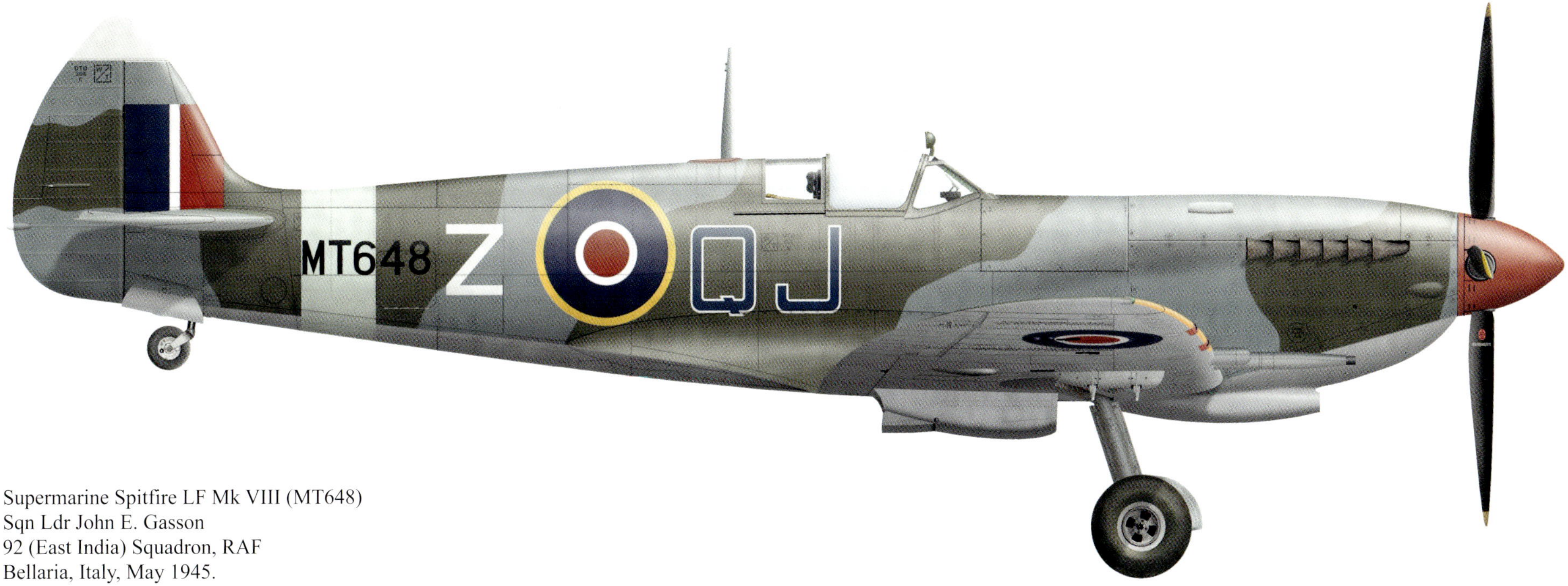

Supermarine Spitfire LF Mk VIII (MT648)
Sqn Ldr John E. Gasson
92 (East India) Squadron, RAF
Bellaria, Italy, May 1945.

Supermarine Spitfire LF Mk VIII (MV483)
Sqn Ldr A. G. Conway
155 Squadron, RAF
Thedaw, Burma, July 1945.

Supermarine Spitfire LF Mk VIII (A58-504)
Fg Off Rex Watson
452 Squadron, RAAF
Balikpapan, Borneo, August 1945.

Supermarine Spitfire LF Mk VIII (MT841)
Plt Off Pat Callaghan
2 Squadron, Royal Indian AF
Kohat, India, spring 1946.

Spitfire Mk IX

Supermarine Spitfire Mk IX (BR624)
Fg Off Michael Donnet
64 Squadron, RAF
Hornchurch, UK, August 1942.

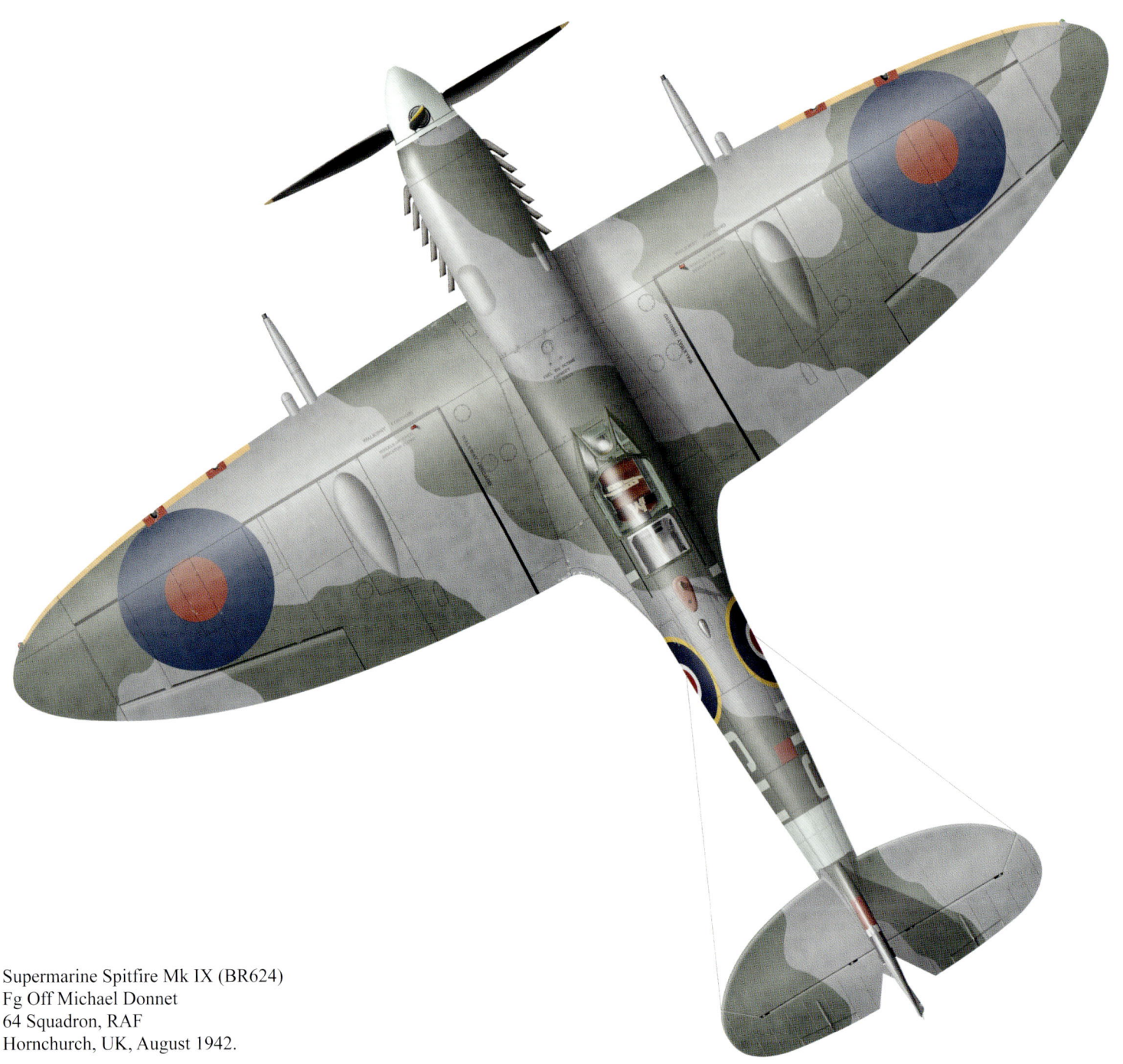

Supermarine Spitfire Mk IX (BR624)
Fg Off Michael Donnet
64 Squadron, RAF
Hornchurch, UK, August 1942.

Supermarine Spitfire Mk IX (BR624)
Fg Off Michael Donnet
64 Squadron, RAF
Hornchurch, UK, August 1942.

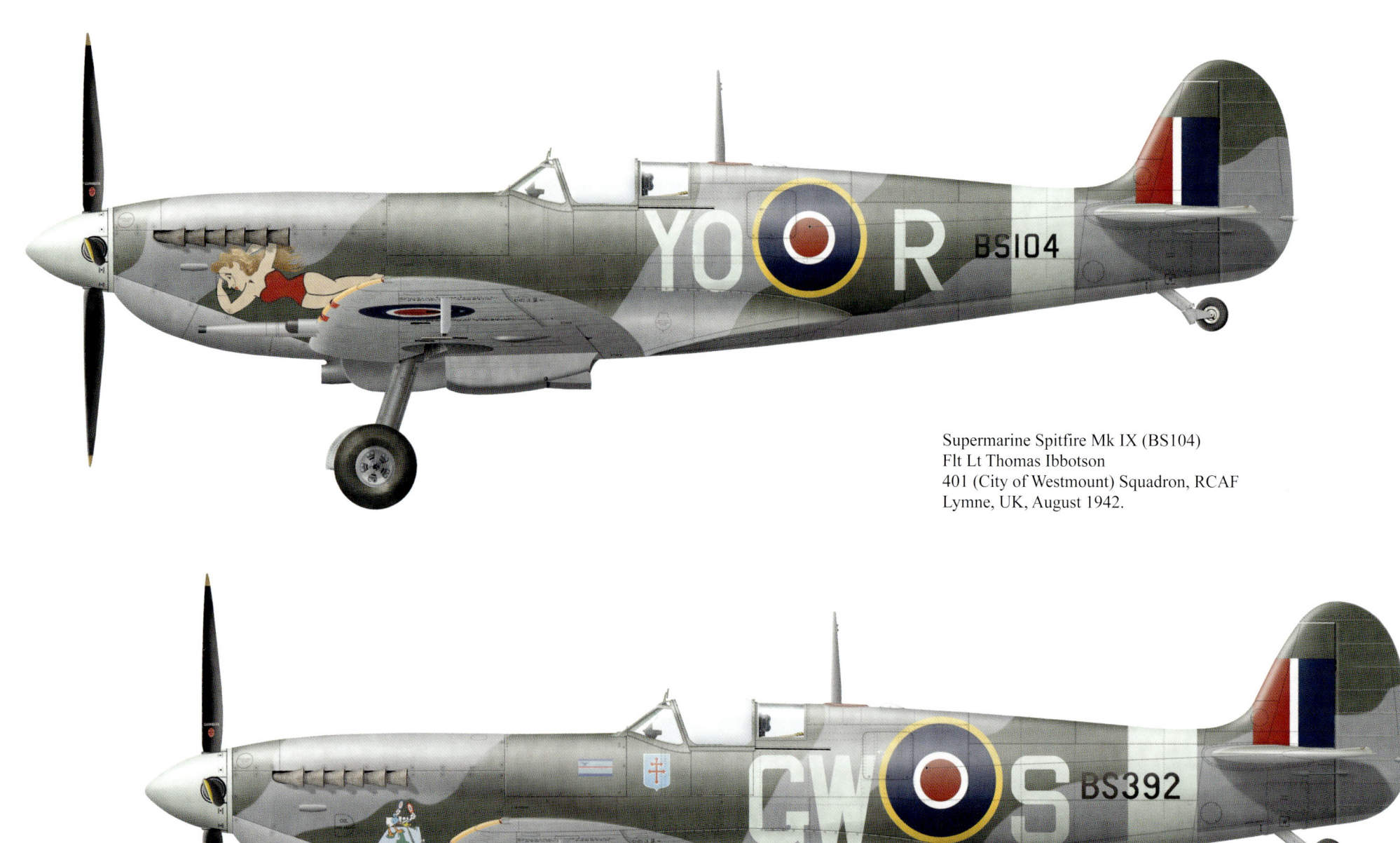

Supermarine Spitfire Mk IX (BS104)
Flt Lt Thomas Ibbotson
401 (City of Westmount) Squadron, RCAF
Lymne, UK, August 1942.

Supermarine Spitfire Mk IX (BS392)
Sqn Ldr Bernard Dupérier
340 (Free French) Squadron, RAF
Biggin Hill, UK, autumn 1942.

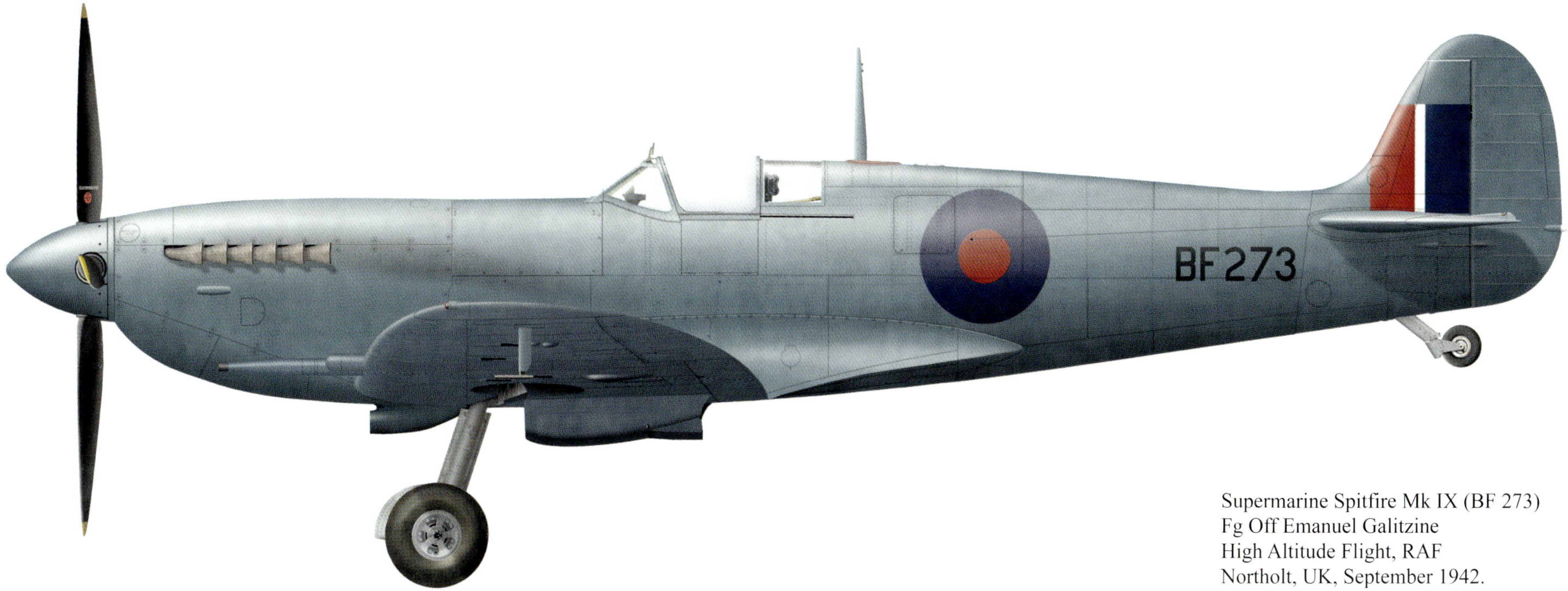

Supermarine Spitfire Mk IX (BF 273)
Fg Off Emanuel Galitzine
High Altitude Flight, RAF
Northolt, UK, September 1942.

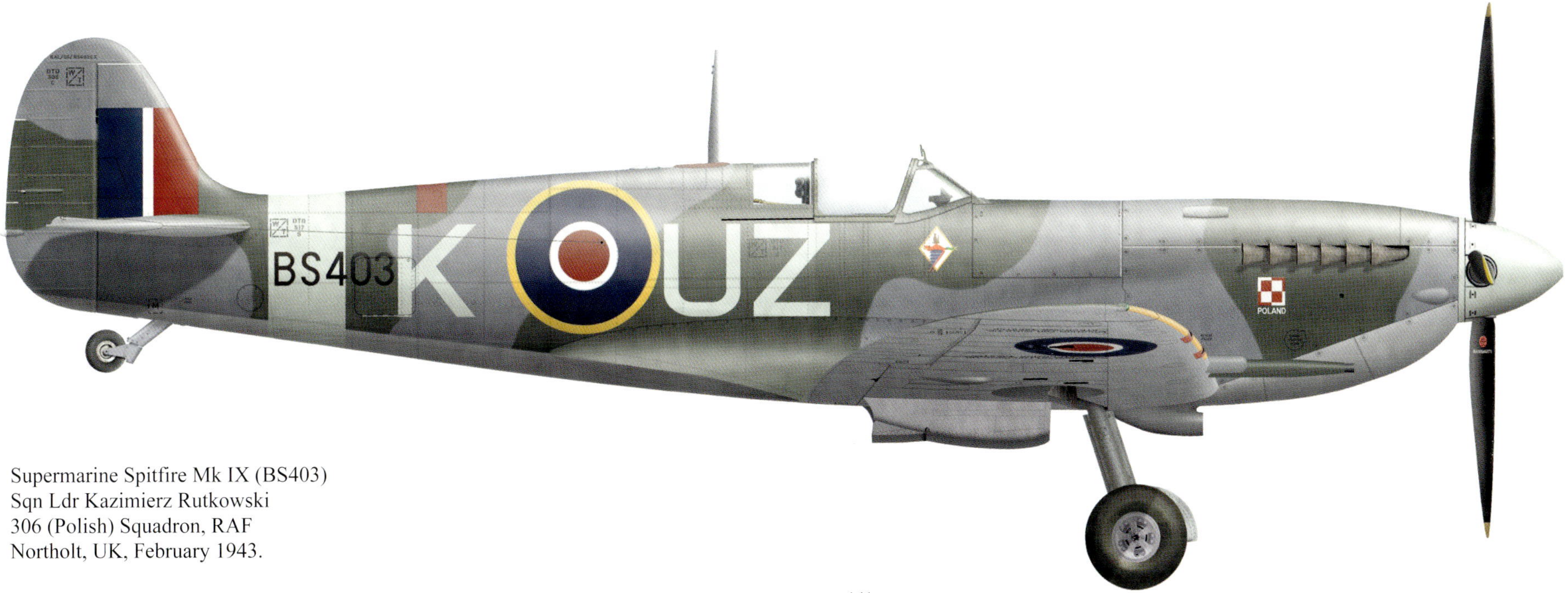

Supermarine Spitfire Mk IX (BS403)
Sqn Ldr Kazimierz Rutkowski
306 (Polish) Squadron, RAF
Northolt, UK, February 1943.

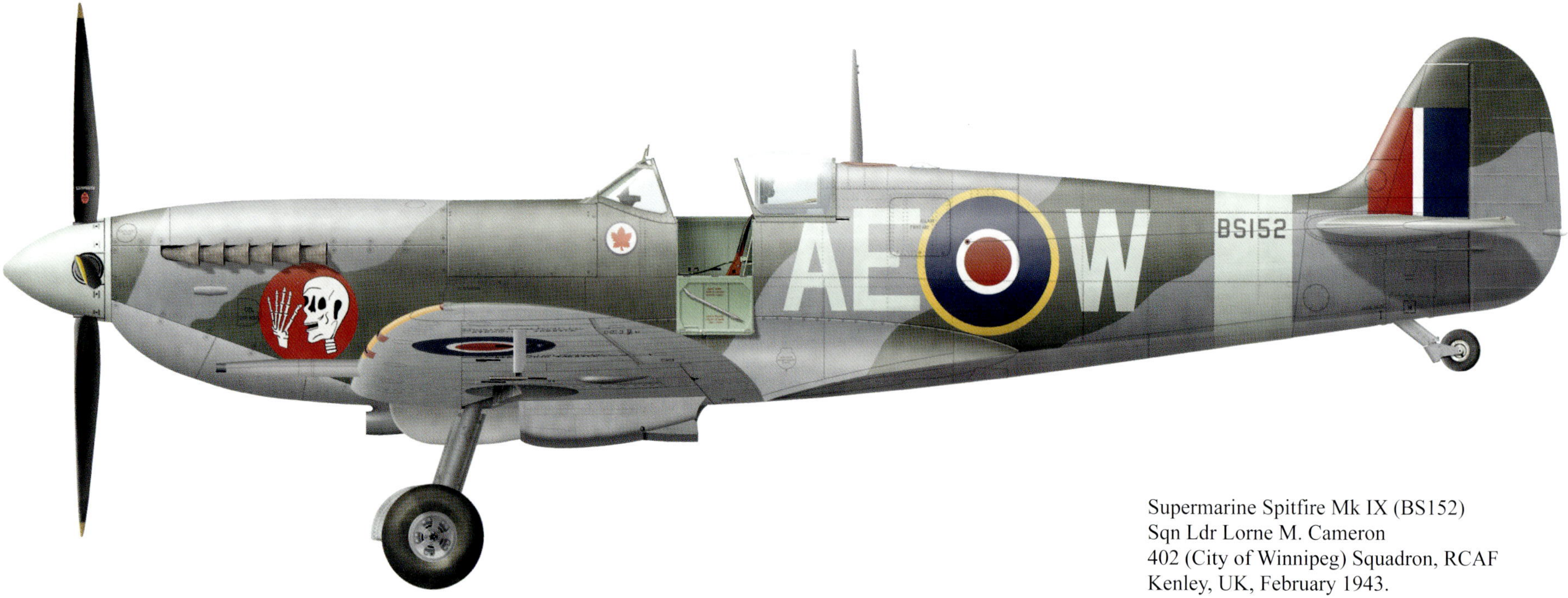

Supermarine Spitfire Mk IX (BS152)
Sqn Ldr Lorne M. Cameron
402 (City of Winnipeg) Squadron, RCAF
Kenley, UK, February 1943.

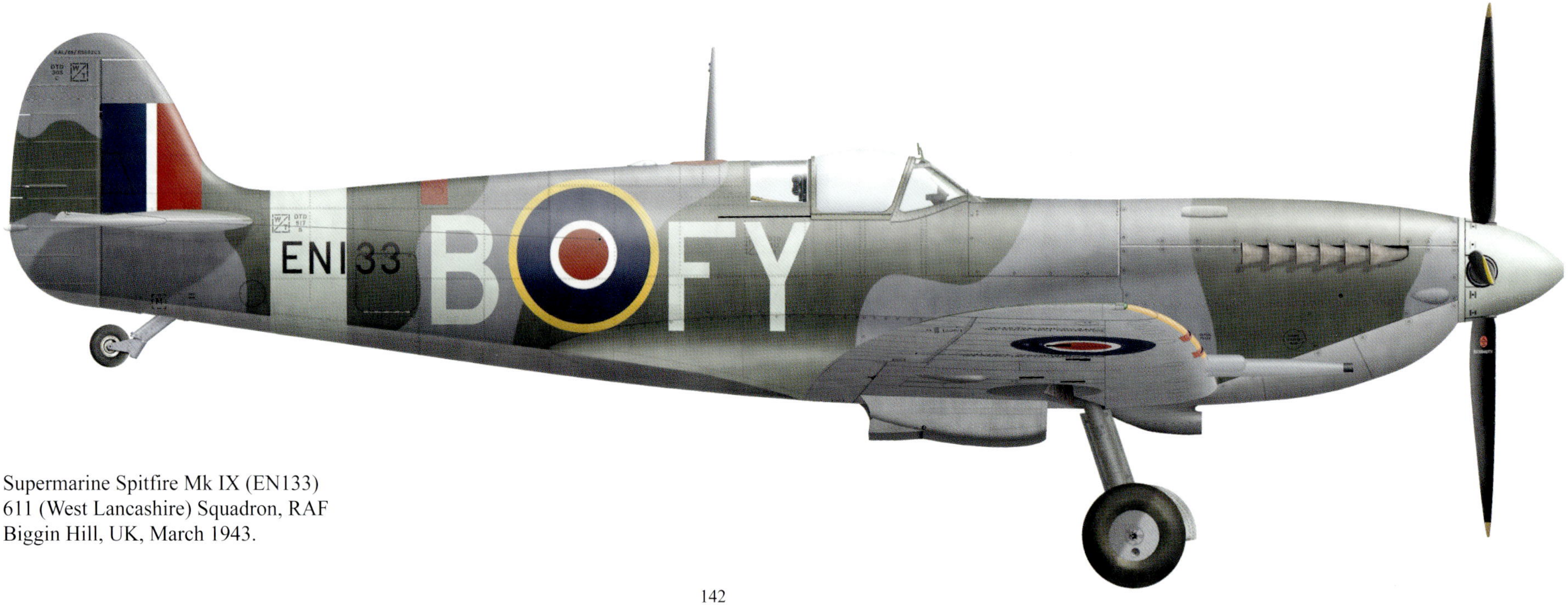

Supermarine Spitfire Mk IX (EN133)
611 (West Lancashire) Squadron, RAF
Biggin Hill, UK, March 1943.

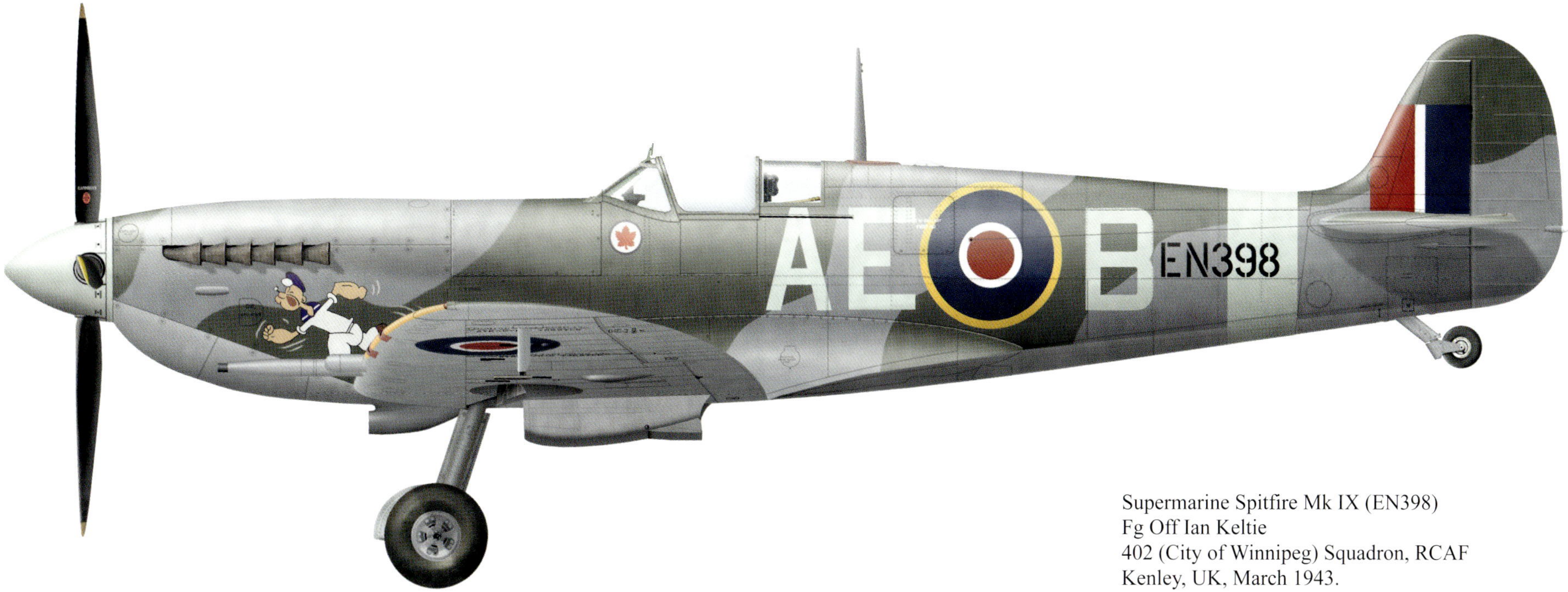

Supermarine Spitfire Mk IX (EN398)
Fg Off Ian Keltie
402 (City of Winnipeg) Squadron, RCAF
Kenley, UK, March 1943.

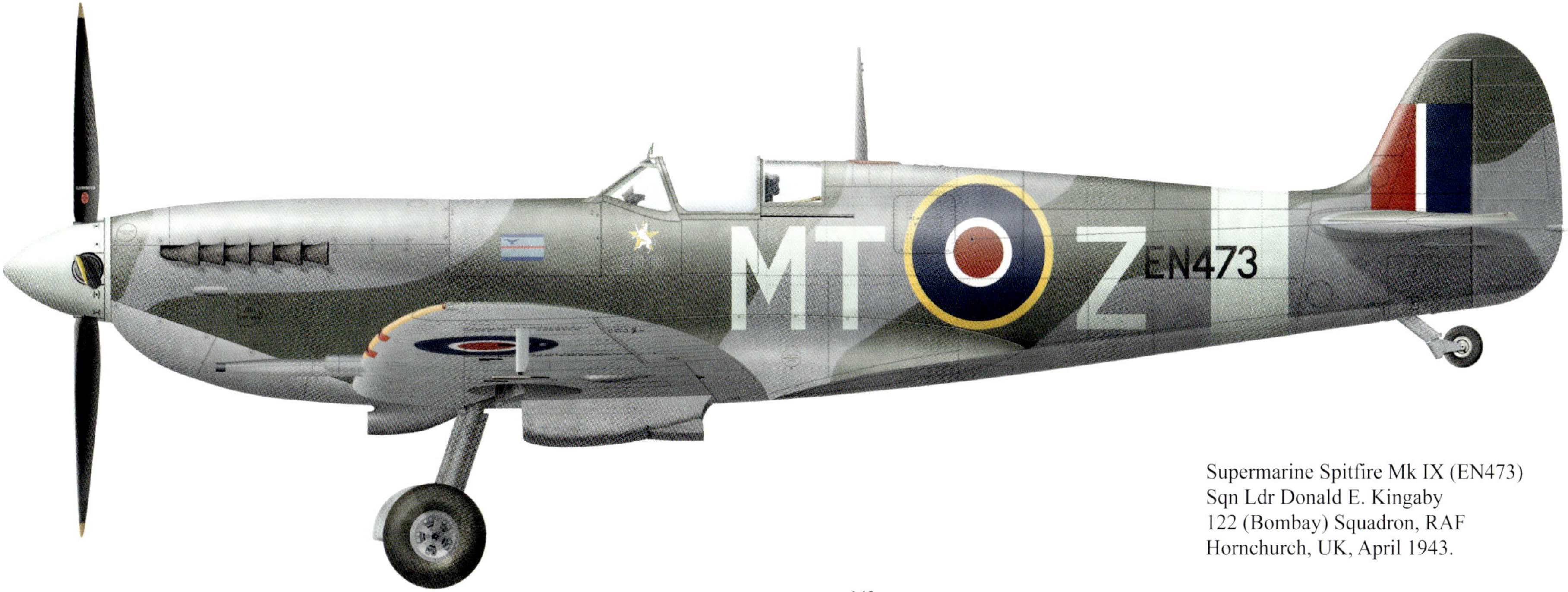

Supermarine Spitfire Mk IX (EN473)
Sqn Ldr Donald E. Kingaby
122 (Bombay) Squadron, RAF
Hornchurch, UK, April 1943.

Supermarine Spitfire Mk IX (BS538)
Sgt Pierre Clostermann
341 (Free French) Squadron, RAF
Biggin Hill, UK, April 1943.

Supermarine Spitfire Mk IX (EN522)
Sqn Ldr John Ratten
453 Squadron, RAAF
Hornchurch, UK, April 1943.

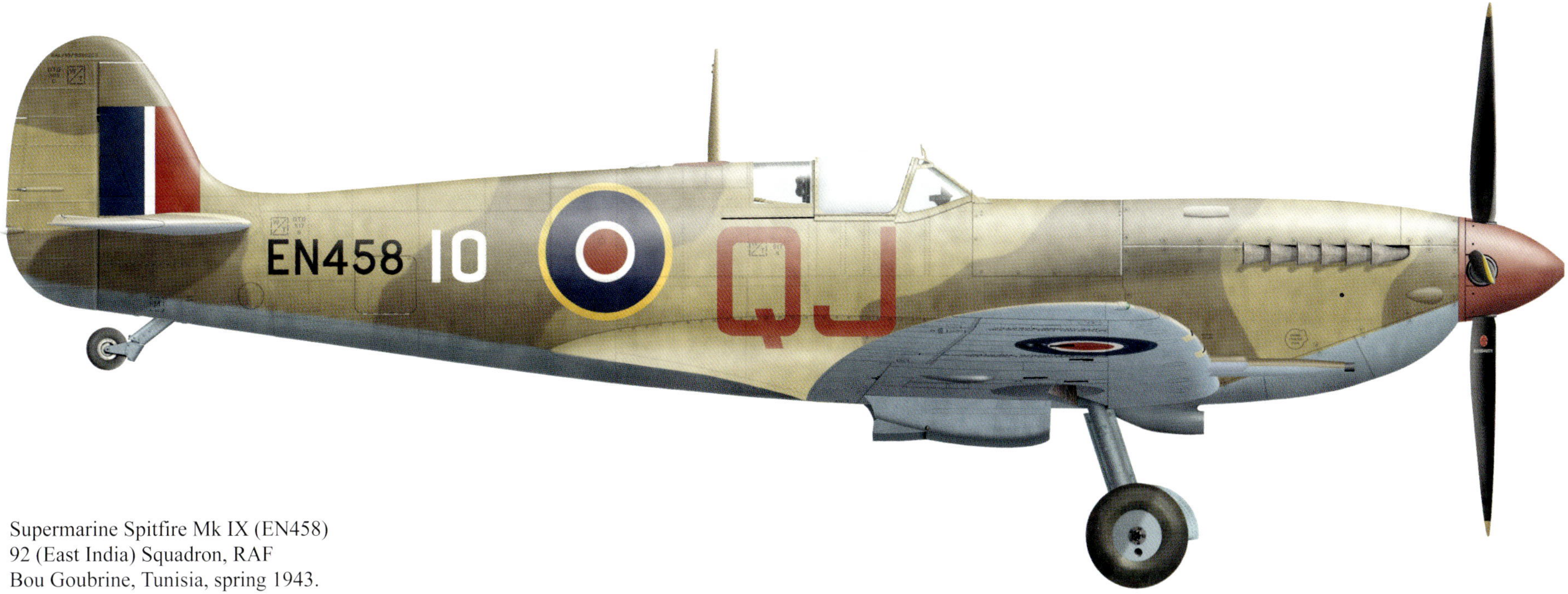

Supermarine Spitfire Mk IX (EN458)
92 (East India) Squadron, RAF
Bou Goubrine, Tunisia, spring 1943.

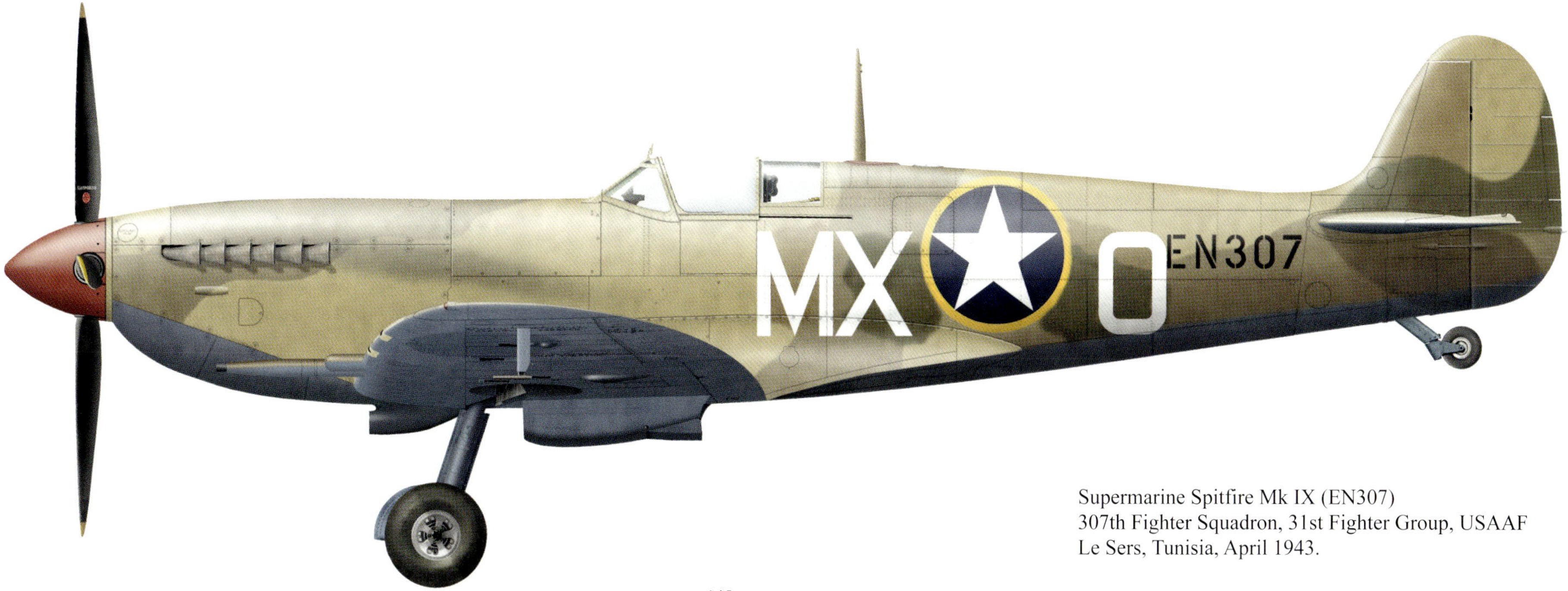

Supermarine Spitfire Mk IX (EN307)
307th Fighter Squadron, 31st Fighter Group, USAAF
Le Sers, Tunisia, April 1943.

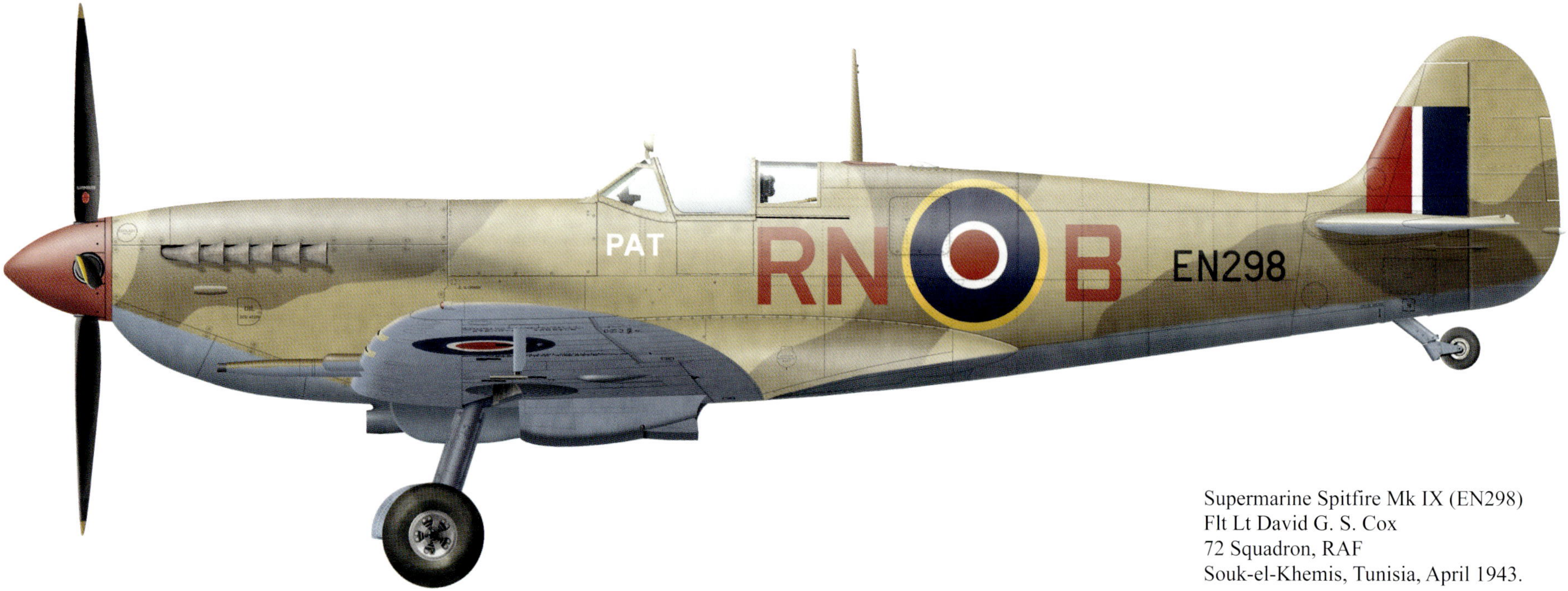

Supermarine Spitfire Mk IX (EN298)
Flt Lt David G. S. Cox
72 Squadron, RAF
Souk-el-Khemis, Tunisia, April 1943.

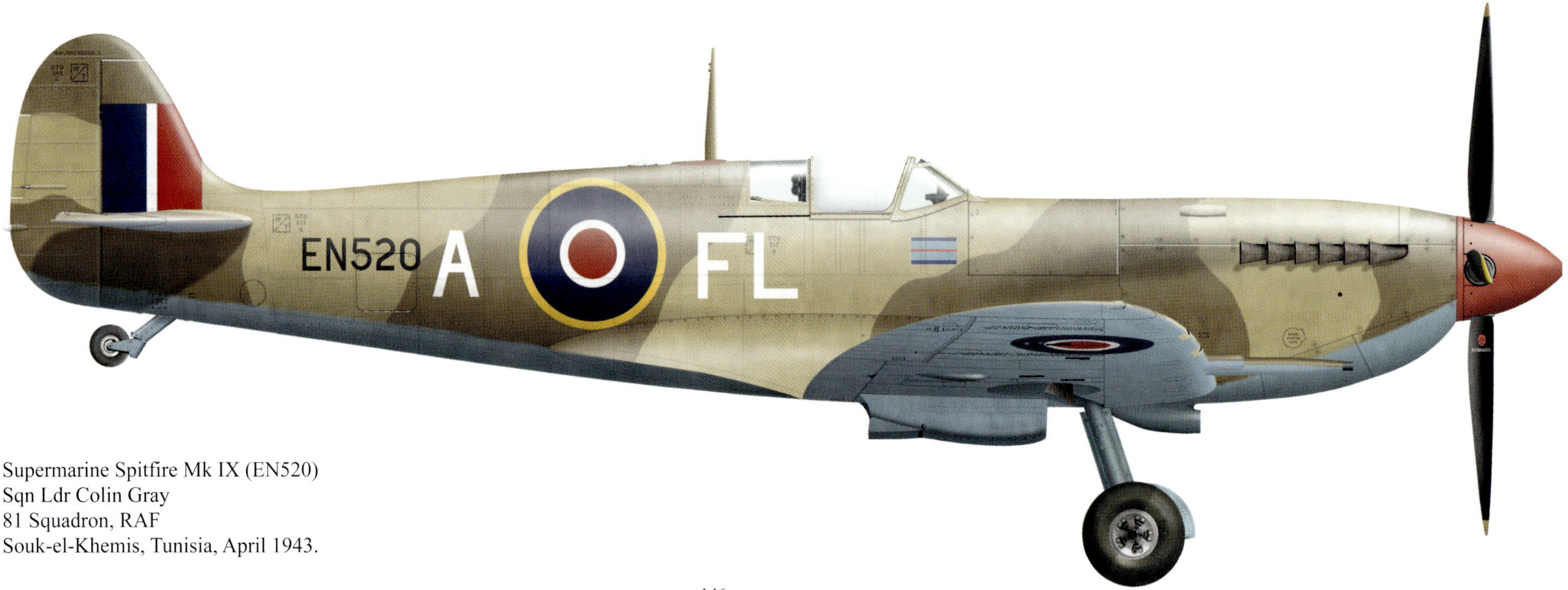

Supermarine Spitfire Mk IX (EN520)
Sqn Ldr Colin Gray
81 Squadron, RAF
Souk-el-Khemis, Tunisia, April 1943.

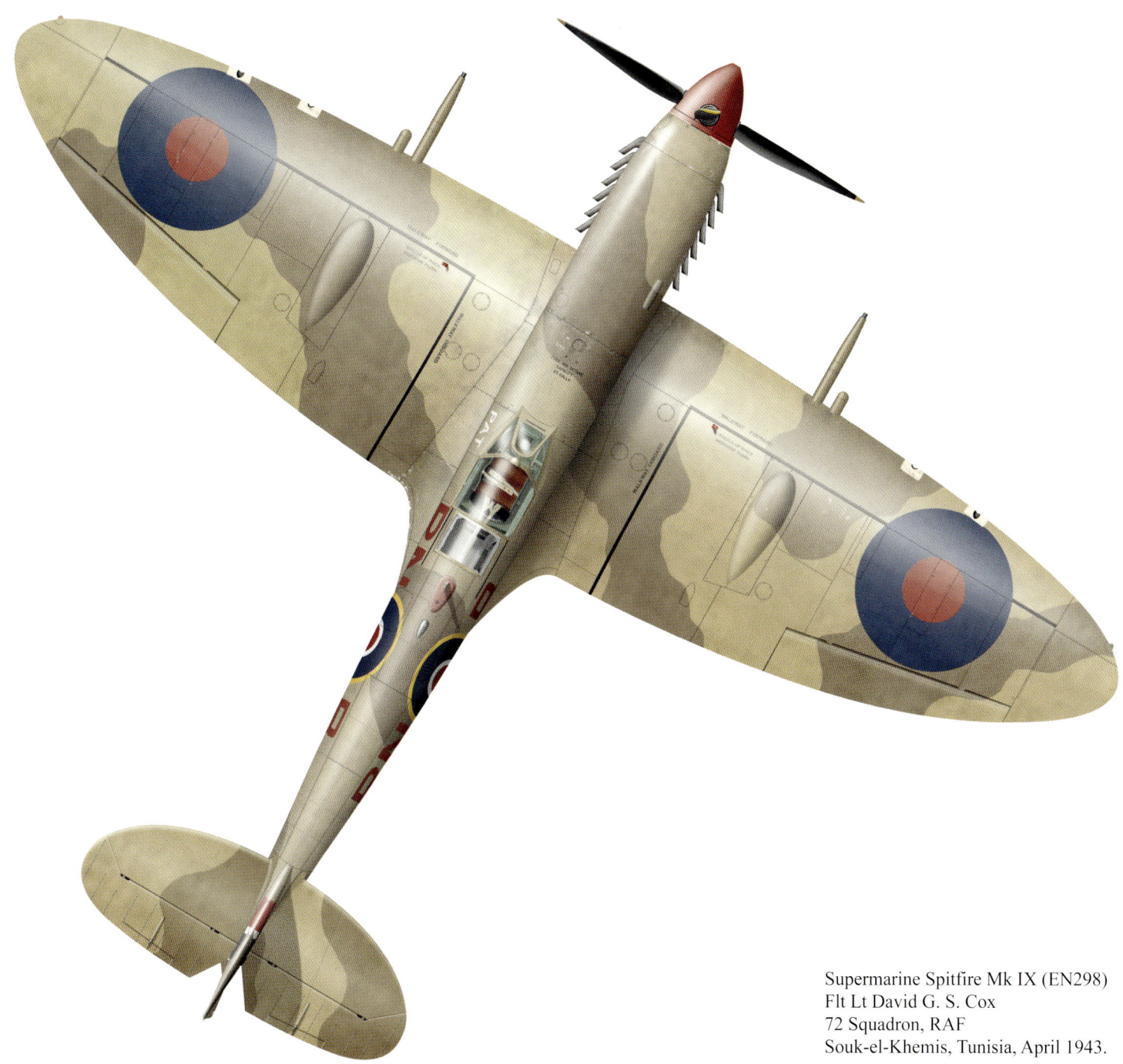

Supermarine Spitfire Mk IX (EN298)
Flt Lt David G. S. Cox
72 Squadron, RAF
Souk-el-Khemis, Tunisia, April 1943.

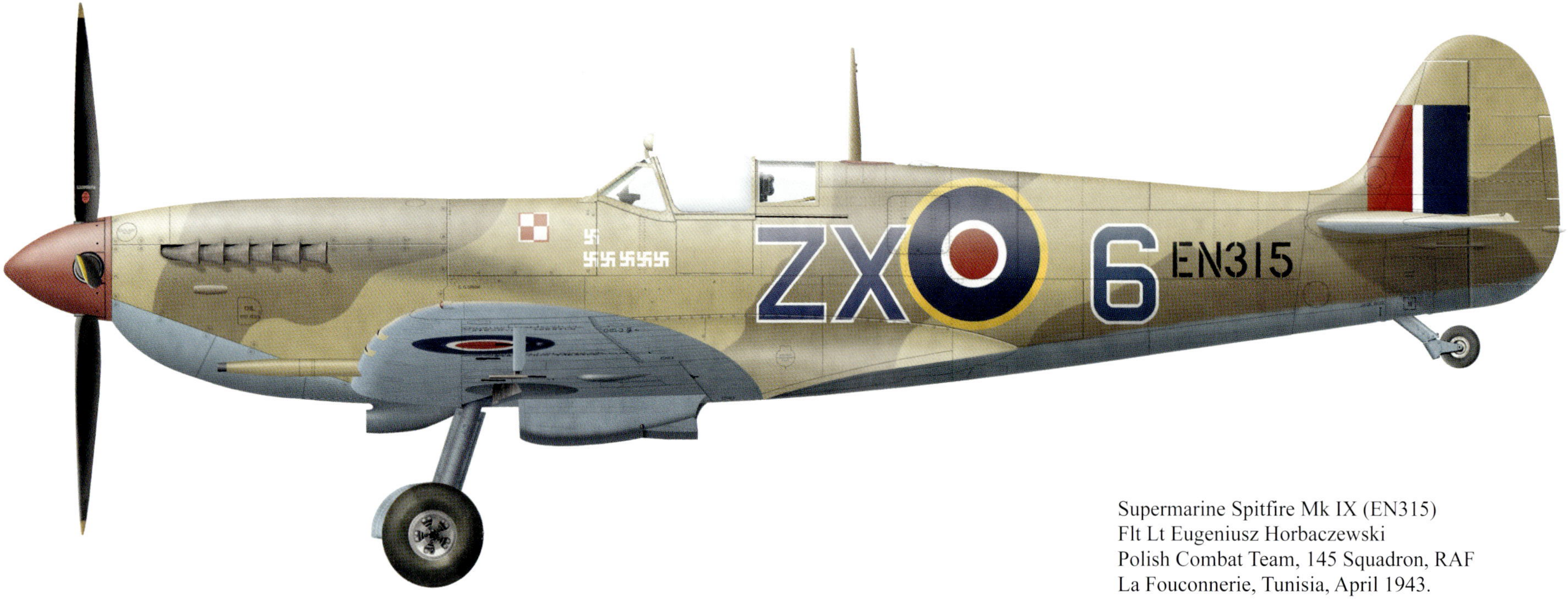

Supermarine Spitfire Mk IX (EN315)
Flt Lt Eugeniusz Horbaczewski
Polish Combat Team, 145 Squadron, RAF
La Fouconnerie, Tunisia, April 1943.

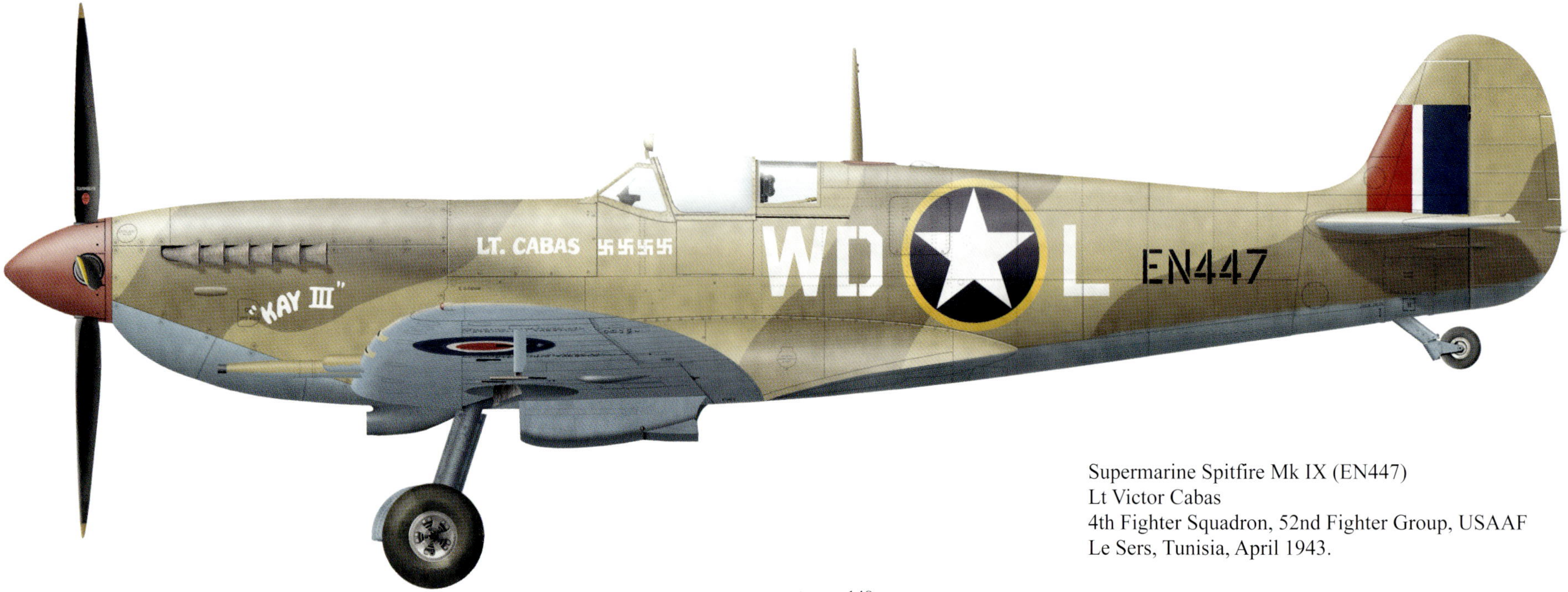

Supermarine Spitfire Mk IX (EN447)
Lt Victor Cabas
4th Fighter Squadron, 52nd Fighter Group, USAAF
Le Sers, Tunisia, April 1943.

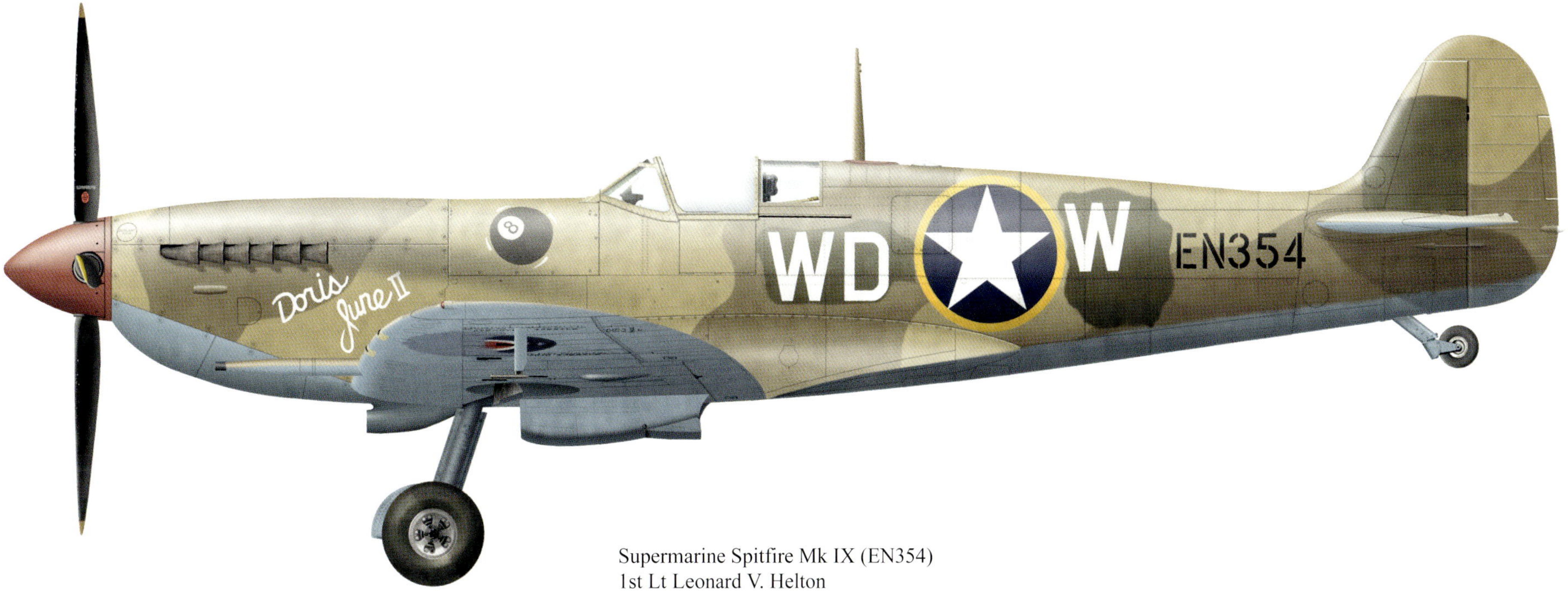

Supermarine Spitfire Mk IX (EN354)
1st Lt Leonard V. Helton
4th Fighter Squadron, 52nd Fighter Group, USAAF
La Sebala, Tunis, June 1943.

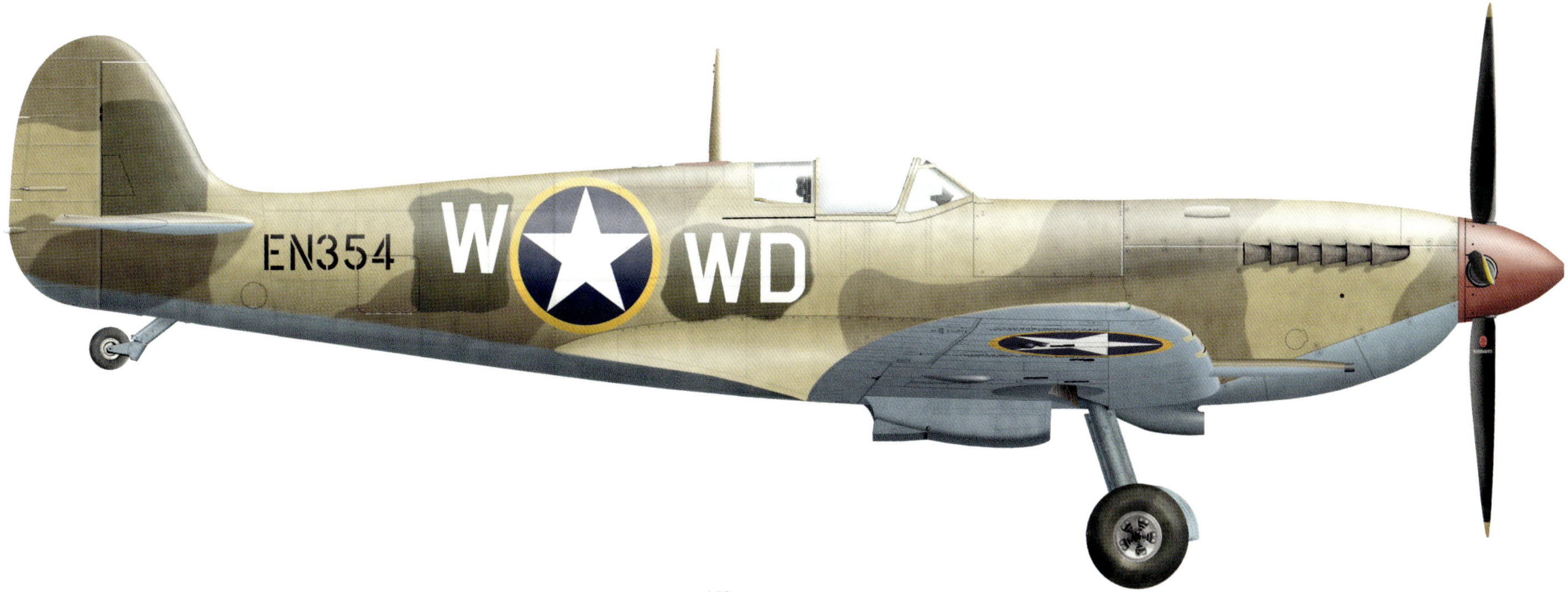

Supermarine Spitfire Mk IX (EN354)
1st Lt Leonard V. Helton
4th Fighter Squadron, 52nd Fighter Group, USAAF
La Sebala, Tunis, June 1943.

Supermarine Spitfire Mk IX (EN354)
1st Lt Leonard V. Helton
4th Fighter Squadron, 52nd Fighter Group, USAAF
La Sebala, Tunis, June 1943.

Supermarine Spitfire Mk IX (EN286)
Flt Lt Eric Robinson
1 Squadron, SAAF
Luqa, Malta, June 1943.

Supermarine Spitfire Mk IX (EN500)
Flt Off Irving F. Kennedy
249 Squadron, RAF
Qrendi, Malta, July 1943.

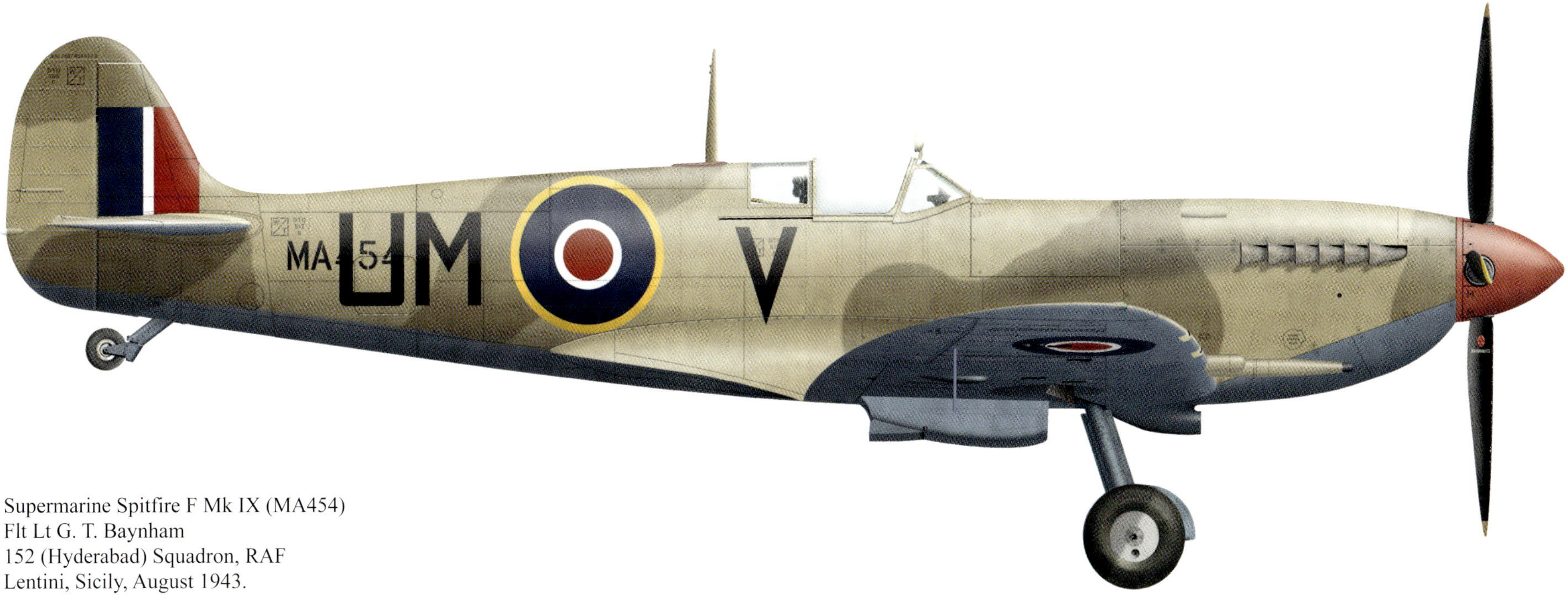

Supermarine Spitfire F Mk IX (MA454)
Flt Lt G. T. Baynham
152 (Hyderabad) Squadron, RAF
Lentini, Sicily, August 1943.

Supermarine Spitfire F Mk IX (serial unknown)
Lt Fred Ohr
2nd Fighter Squadron, 52nd Fighter Group, USAAF
Palermo, Sicily, August 1943.

Supermarine Spitfire F Mk IX (MH358)
Sqn Ldr James E. Storrar
65 (East India) Squadron, RAF
Kingsnorth, UK, September 1943.

Supermarine Spitfire Mk IX (EN526)
Wg Cdr Aleksander Gabszewicz
Northolt Wing, RAF
Northolt, UK, September 1943.

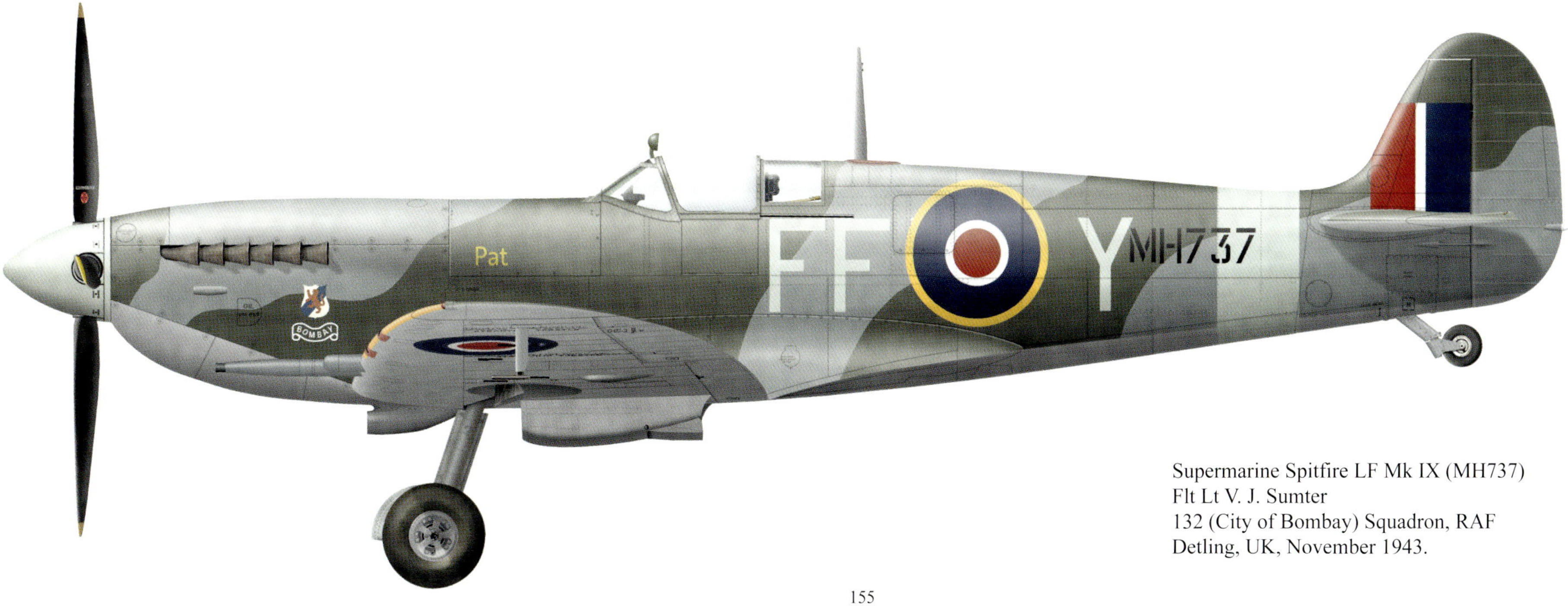

Supermarine Spitfire LF Mk IX (MH737)
Flt Lt V. J. Sumter
132 (City of Bombay) Squadron, RAF
Detling, UK, November 1943.

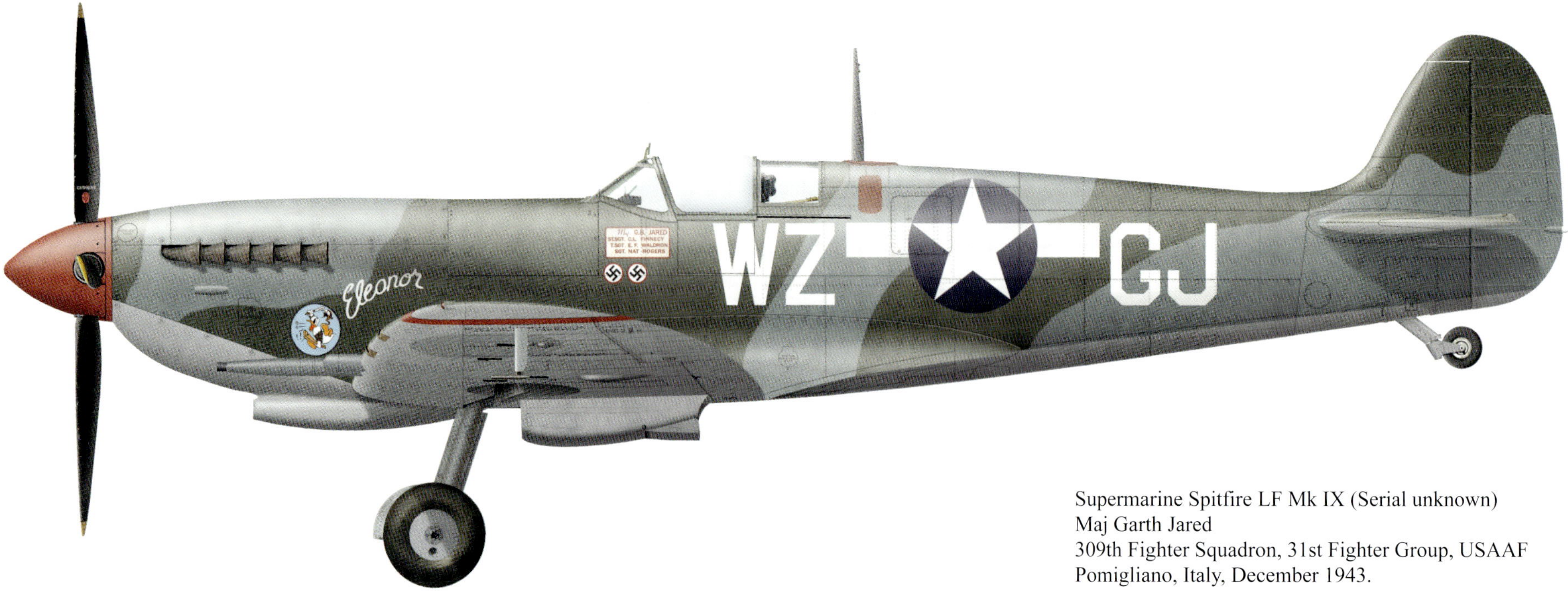

Supermarine Spitfire LF Mk IX (Serial unknown)
Maj Garth Jared
309th Fighter Squadron, 31st Fighter Group, USAAF
Pomigliano, Italy, December 1943.

Supermarine Spitfire LF Mk IX (serial unknown)
Lt Robert Conno
309th Fighter Squadron, 31st Fighter Group, USAAF
Castel Volturno, Italy, February 1944.

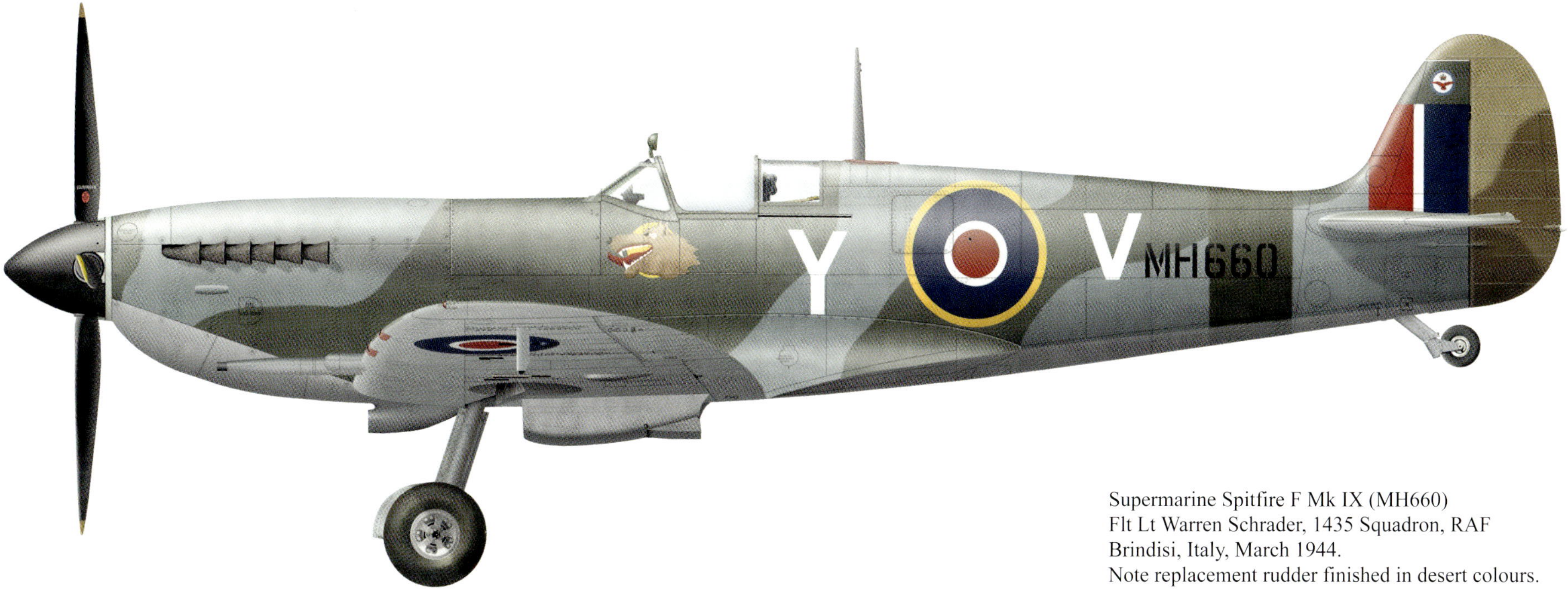

Supermarine Spitfire F Mk IX (MH660)
Flt Lt Warren Schrader, 1435 Squadron, RAF
Brindisi, Italy, March 1944.
Note replacement rudder finished in desert colours.

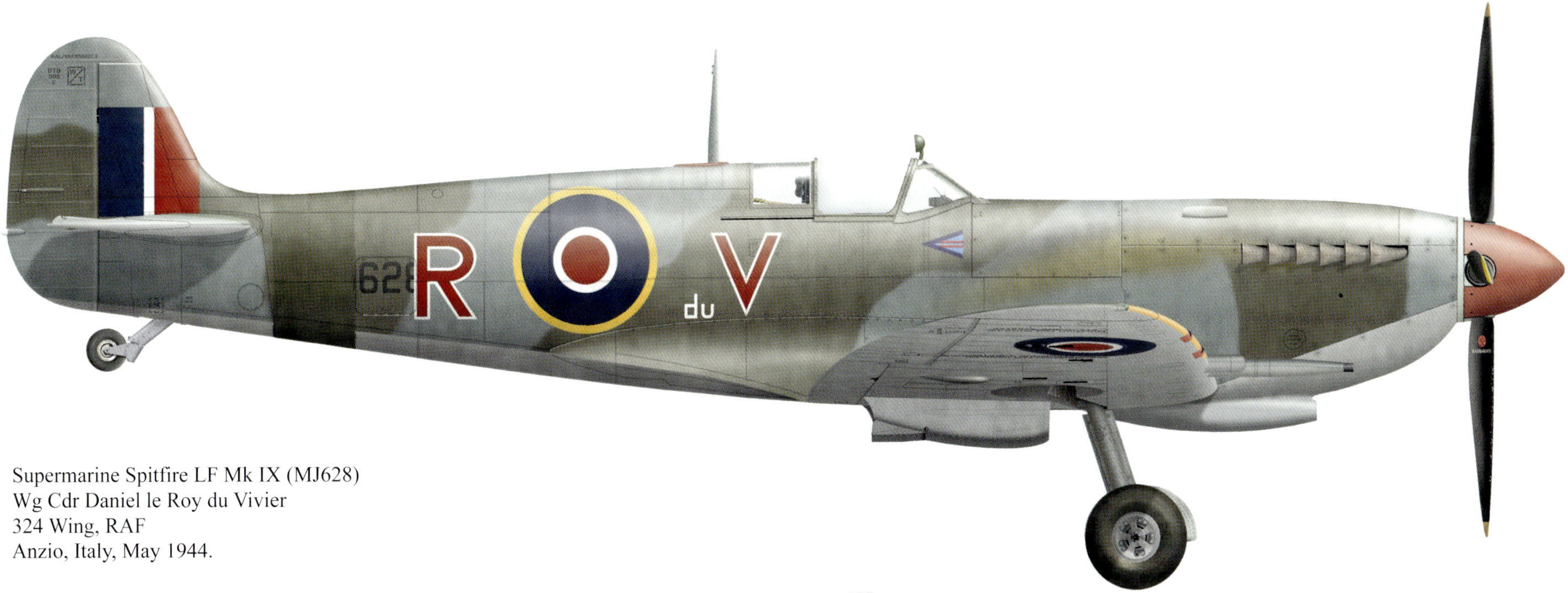

Supermarine Spitfire LF Mk IX (MJ628)
Wg Cdr Daniel le Roy du Vivier
324 Wing, RAF
Anzio, Italy, May 1944.

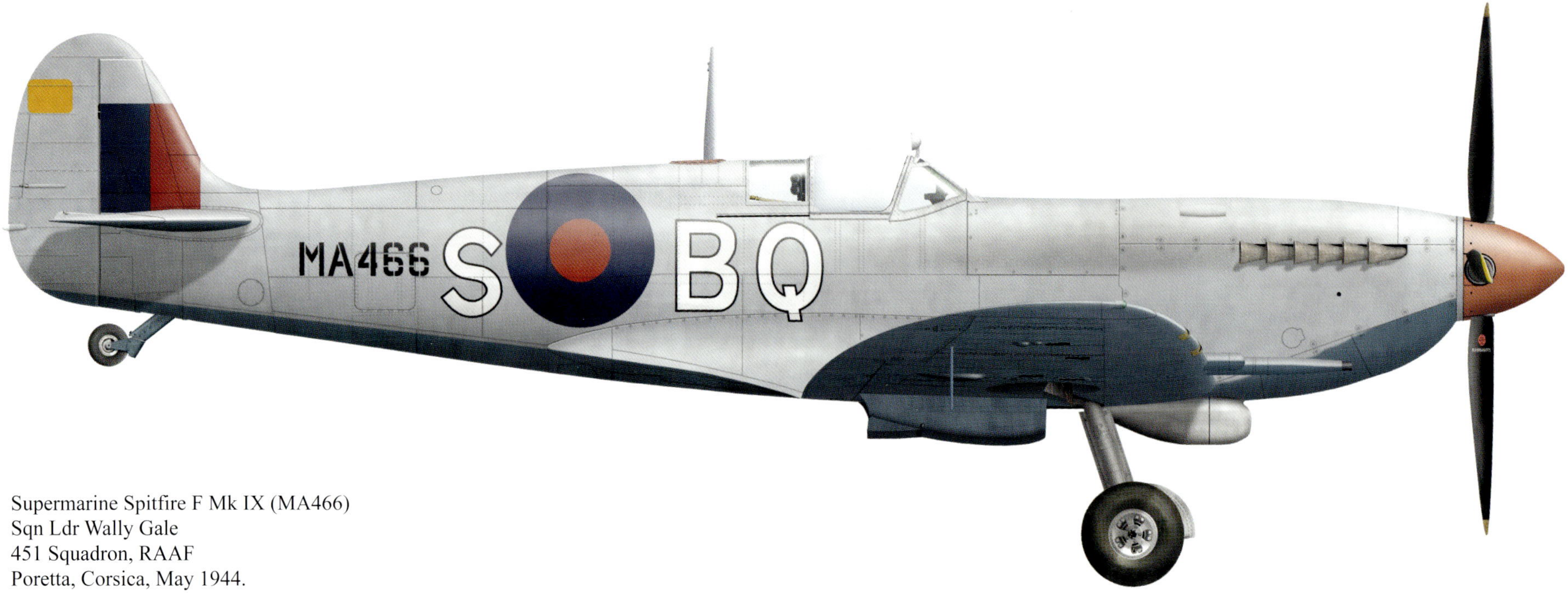

Supermarine Spitfire F Mk IX (MA466)
Sqn Ldr Wally Gale
451 Squadron, RAAF
Poretta, Corsica, May 1944.

Supermarine Spitfire F Mk IX (MA766)
Sqn Ldr Russell Foskett
94 Squadron, RAF
Bu Amud, Egypt, June 1944.

Supermarine Spitfire LF Mk IX (MJ449)
Flt Lt Frantisek Truhlar
312 (Czechoslovak) Squadron, RAF
Appledram, UK, June 1944.

Supermarine Spitfire LF Mk IX (MJ449)
Flt Lt Frantisek Truhlar
312 (Czechoslovak) Squadron, RAF
Appledram, UK, June 1944.

Supermarine Spitfire LF Mk IX (MJ449)
Flt Lt Frantisek Truhlar
312 (Czechoslovak) Squadron, RAF
Appledram, UK, June 1944.

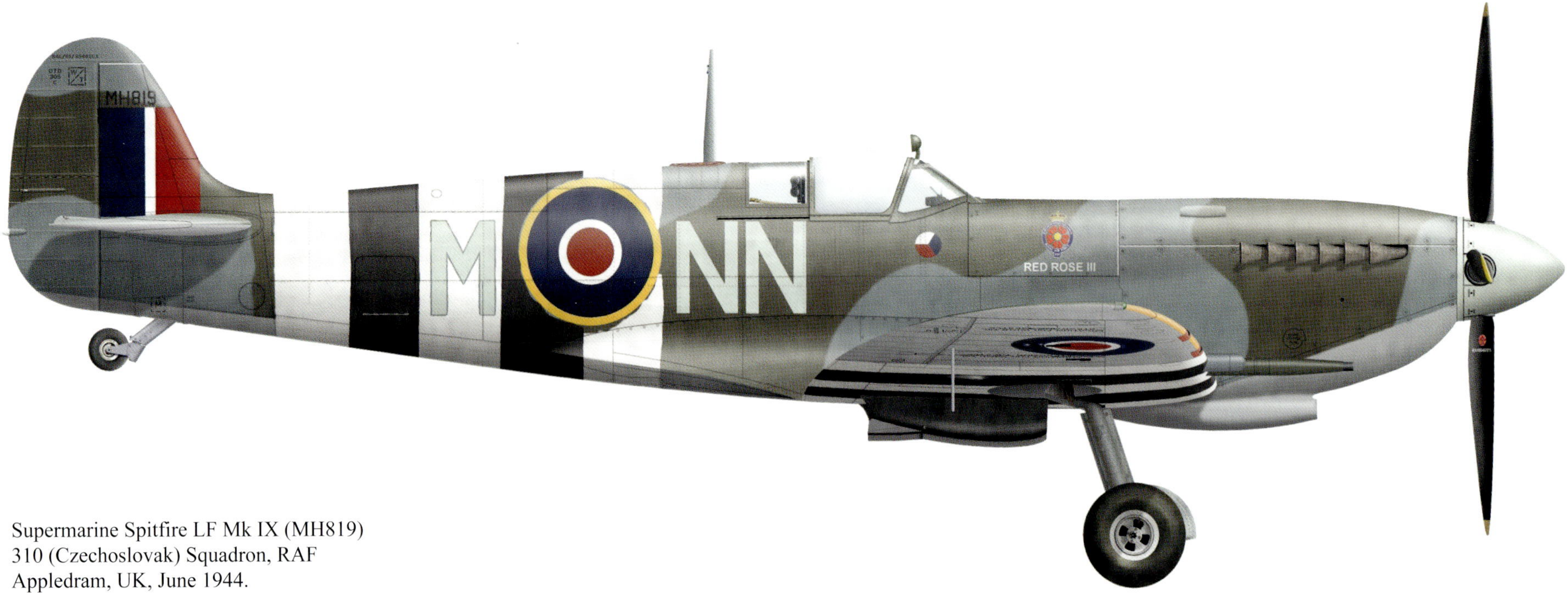

Supermarine Spitfire LF Mk IX (MH819)
310 (Czechoslovak) Squadron, RAF
Appledram, UK, June 1944.

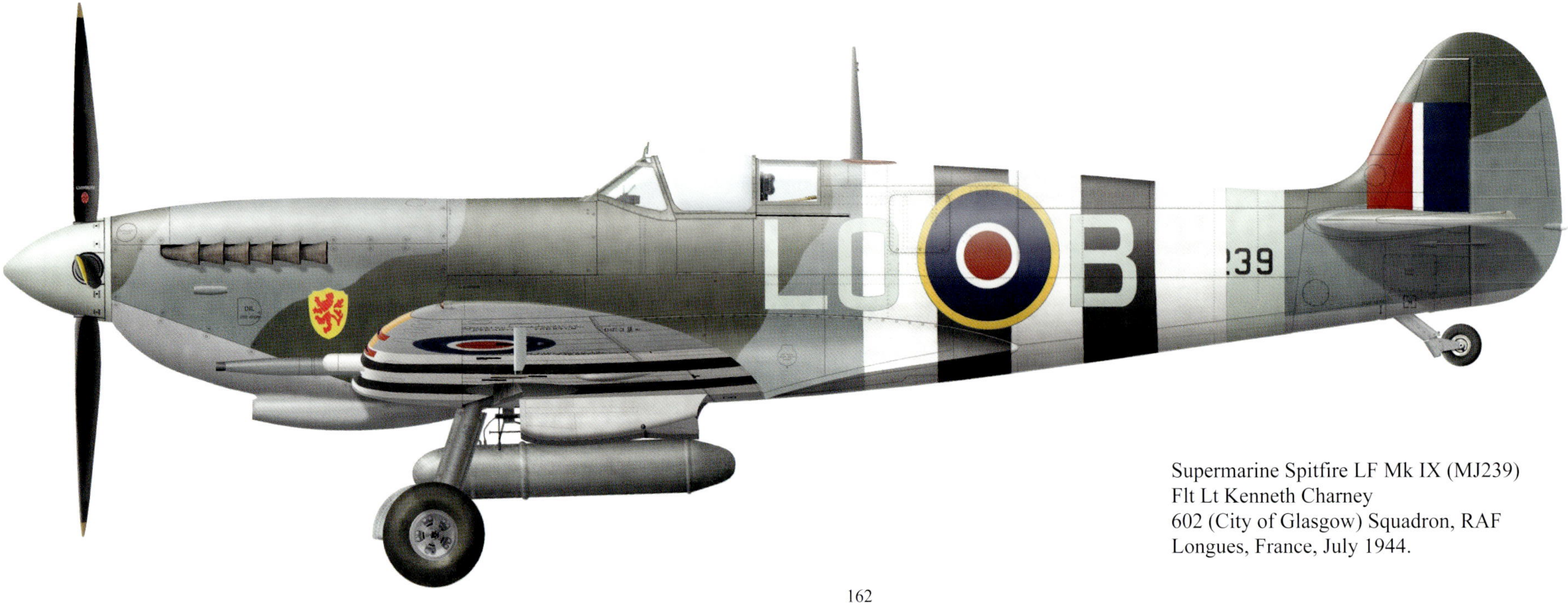

Supermarine Spitfire LF Mk IX (MJ239)
Flt Lt Kenneth Charney
602 (City of Glasgow) Squadron, RAF
Longues, France, July 1944.

Supermarine Spitfire LF Mk IX (MK716)
16 (Reconnaissance) Squadron, RAF
Bayeaux, France, September 1944.
Finished in a special PRU Pink colour for dawn
and dusk low-level sorties in cloudy skies.

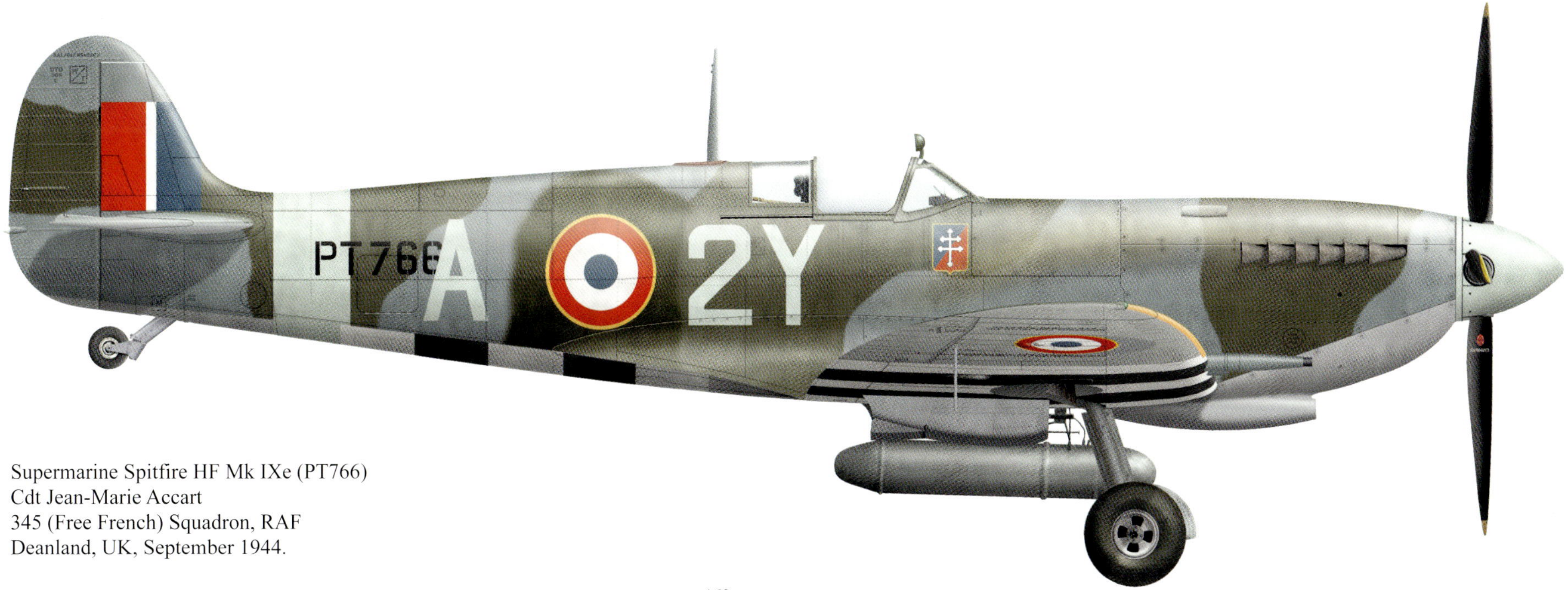

Supermarine Spitfire HF Mk IXe (PT766)
Cdt Jean-Marie Accart
345 (Free French) Squadron, RAF
Deanland, UK, September 1944.

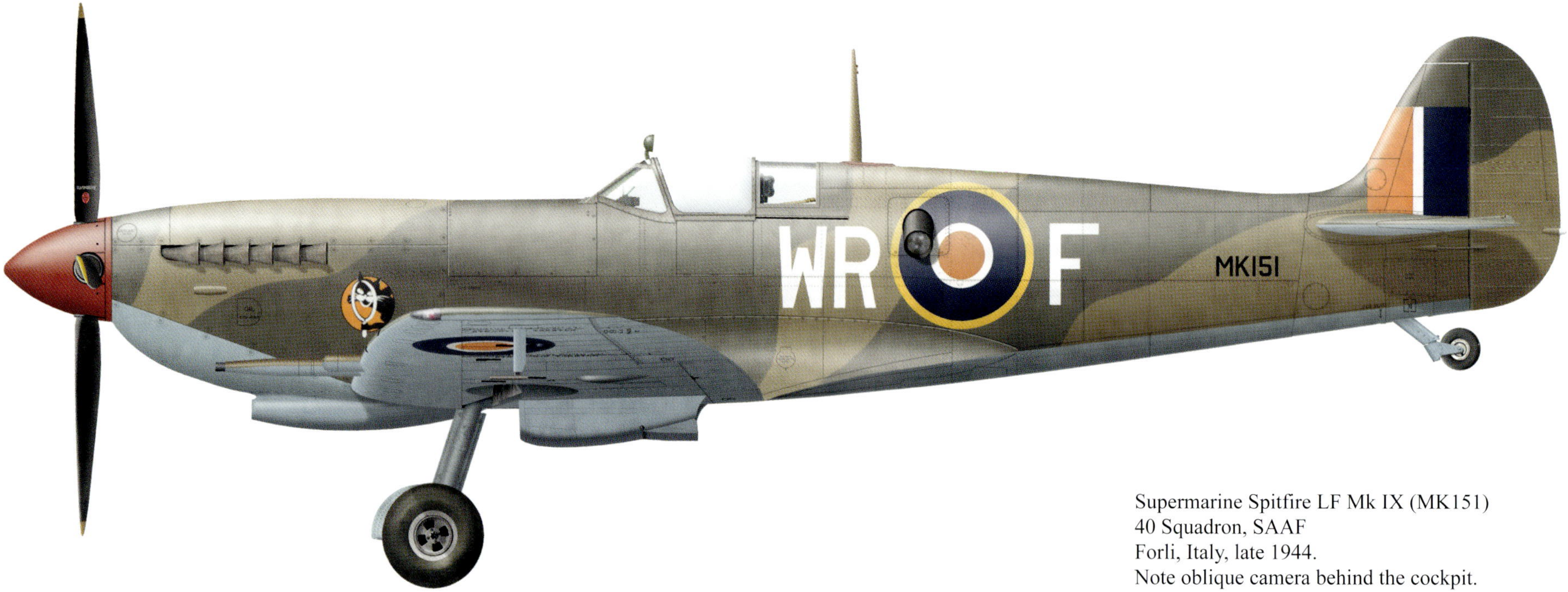

Supermarine Spitfire LF Mk IX (MK151)
40 Squadron, SAAF
Forli, Italy, late 1944.
Note oblique camera behind the cockpit.

Supermarine Spitfire LF Mk IXe (MH712)
WO Henryk Dygala
302 (Polish) Squadron, RAF
St Denijs, Belgium, late 1944.

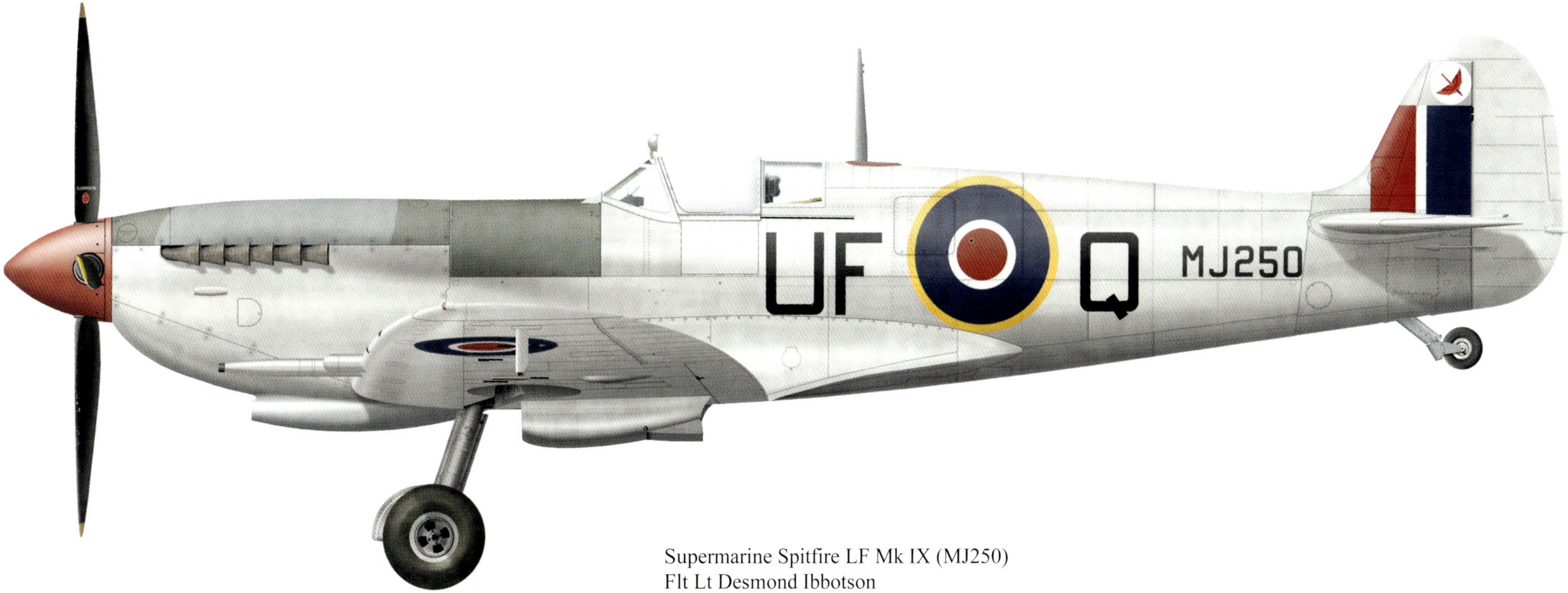

Supermarine Spitfire LF Mk IX (MJ250)
Flt Lt Desmond Ibbotson
601 (County of London) Squadron, RAF
Fano, Italy, November 1944.

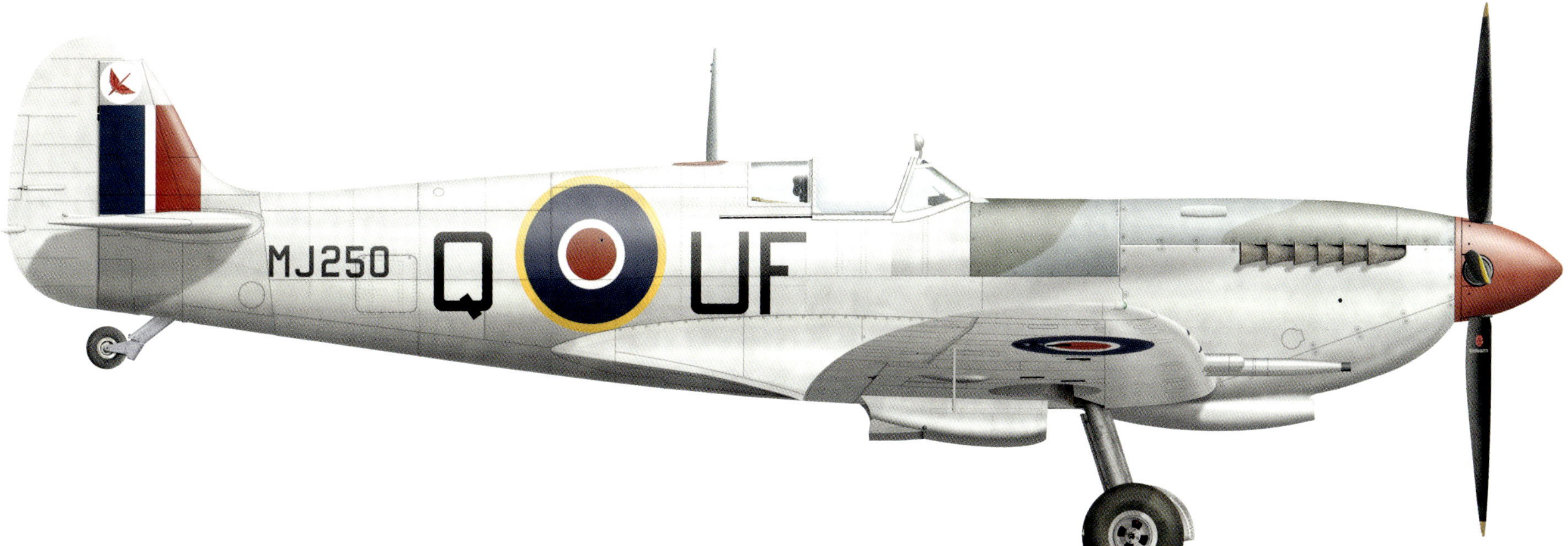

Supermarine Spitfire LF Mk IX (MJ250)
Flt Lt Desmond Ibbotson
601 (County of London) Squadron, RAF
Fano, Italy, November 1944.

Supermarine Spitfire LF Mk IX (MJ250)
Flt Lt Desmond Ibbotson
601 (County of London) Squadron, RAF
Fano, Italy, November 1944.

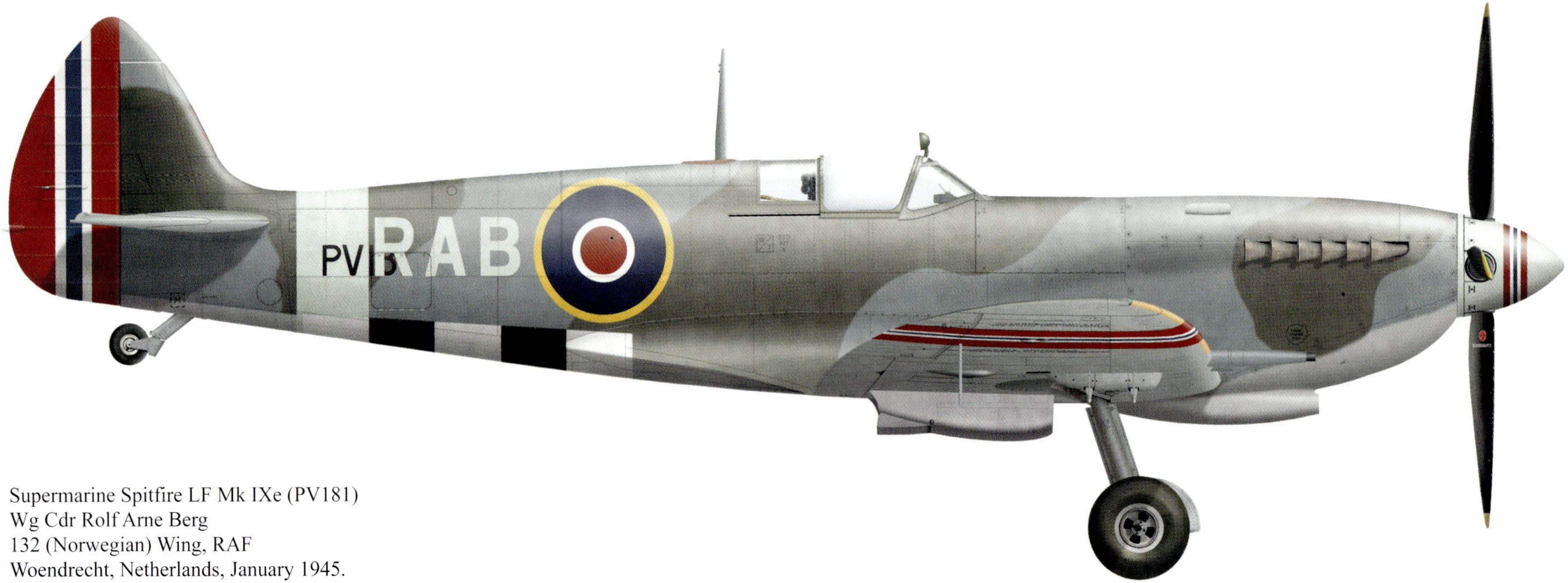

Supermarine Spitfire LF Mk IXe (PV181)
Wg Cdr Rolf Arne Berg
132 (Norwegian) Wing, RAF
Woendrecht, Netherlands, January 1945.

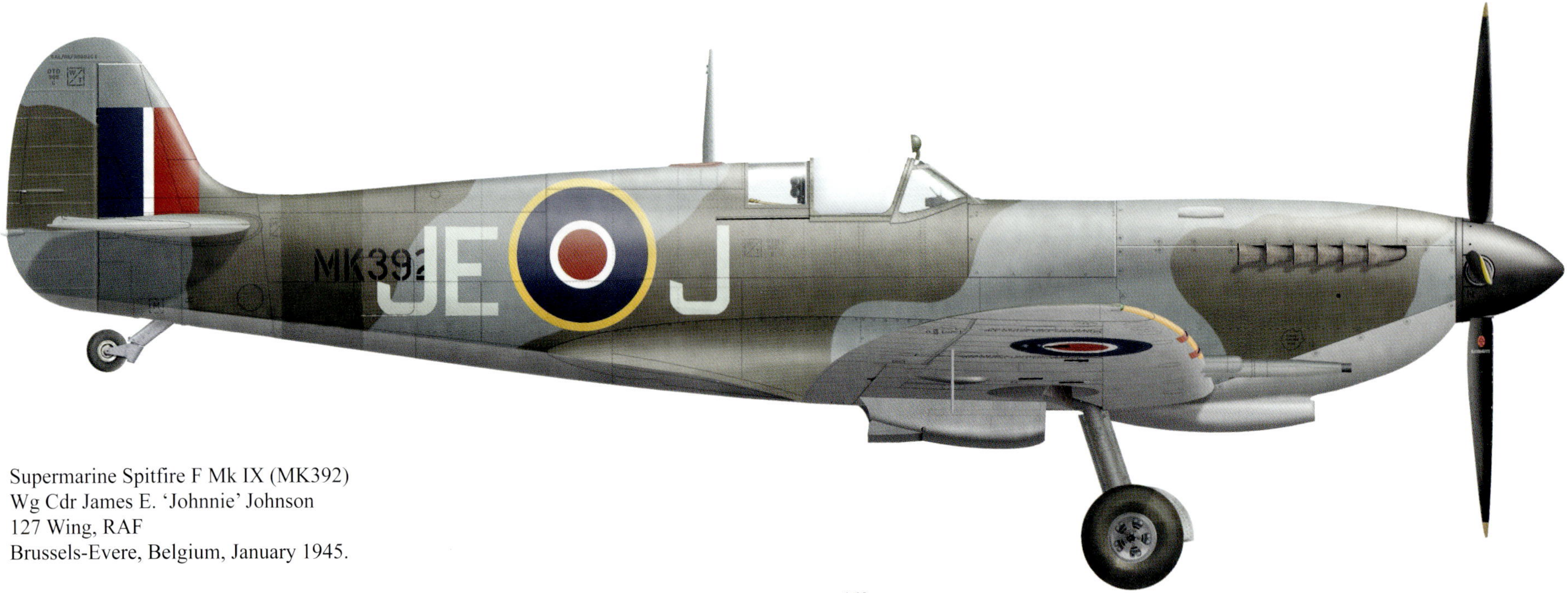

Supermarine Spitfire F Mk IX (MK392)
Wg Cdr James E. 'Johnnie' Johnson
127 Wing, RAF
Brussels-Evere, Belgium, January 1945.

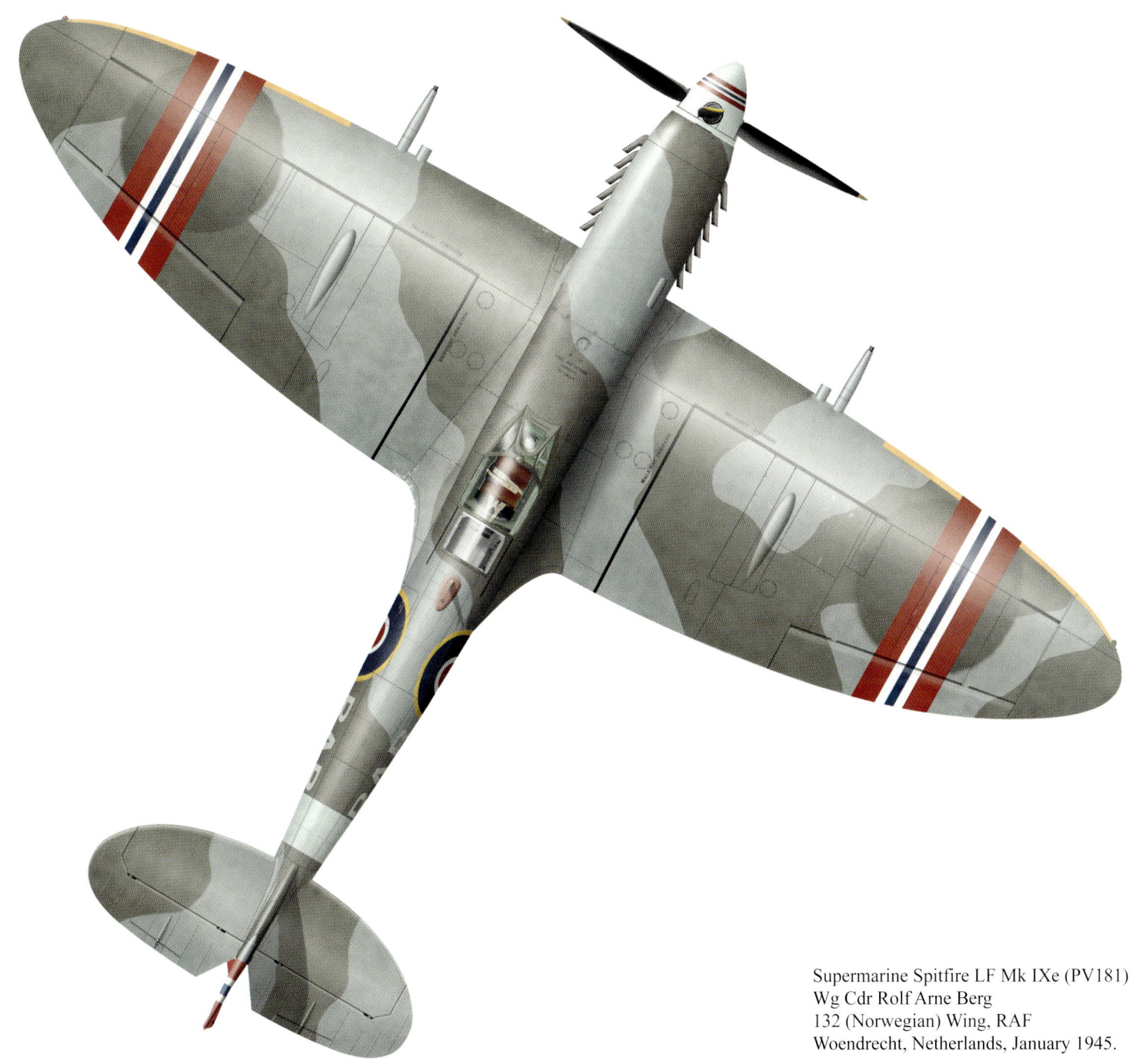

Supermarine Spitfire LF Mk IXe (PV181)
Wg Cdr Rolf Arne Berg
132 (Norwegian) Wing, RAF
Woendrecht, Netherlands, January 1945.

Supermarine Spitfire LF Mk IXe (TA864)
Plt Off Zdzislaw Uchwat
318 (Polish) Squadron, RAF
Risano, Italy, May 1945.

Supermarine Spitfire LF Mk IXe (SM147)
73 Squadron, RAF
Prkos, Yugoslavia, April 1945.

Supermarine Spitfire LF Mk IXe (TA864)
Plt Off Zdzislaw Uchwat
318 (Polish) Squadron, RAF
Risano, Italy, May 1945.

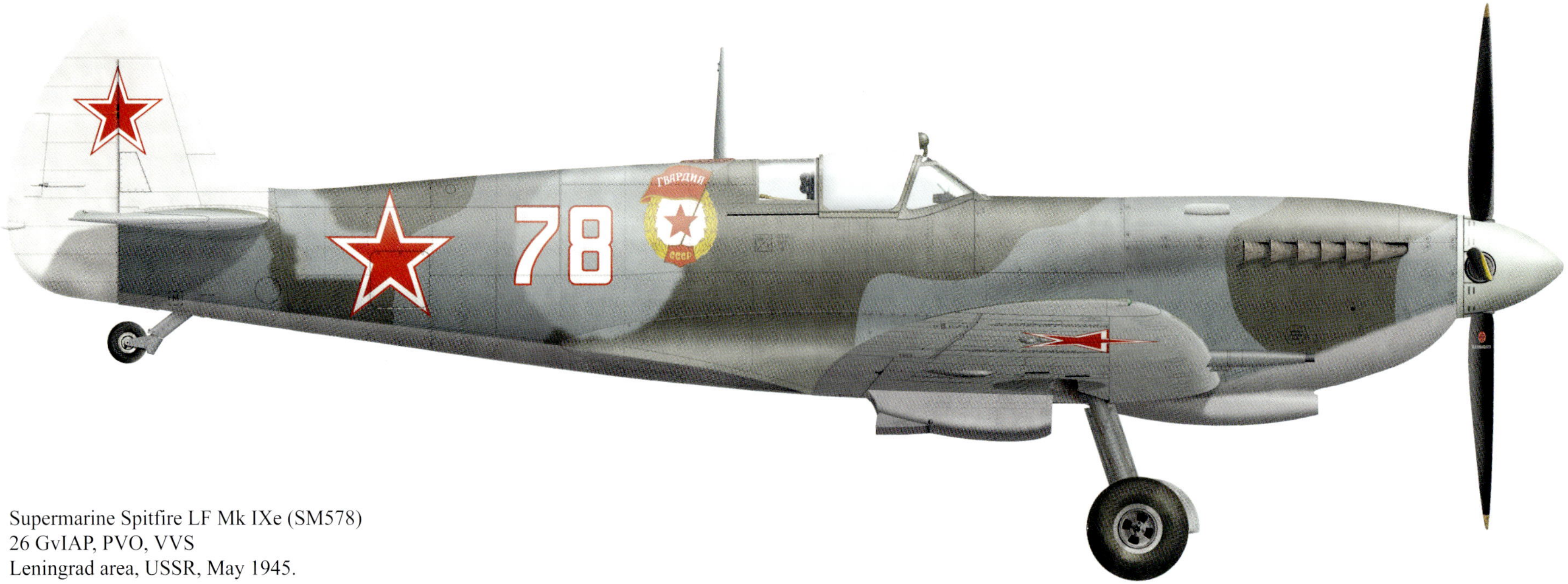

Supermarine Spitfire LF Mk IXe (SM578)
26 GvIAP, PVO, VVS
Leningrad area, USSR, May 1945.

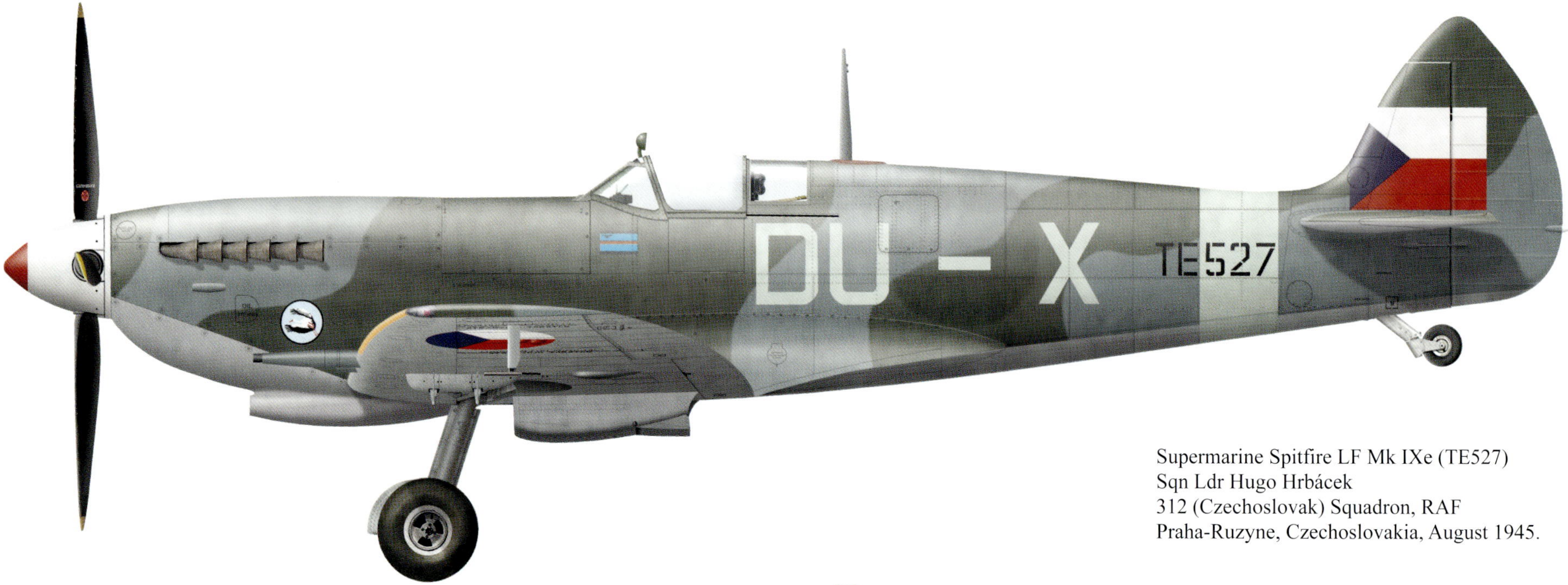

Supermarine Spitfire LF Mk IXe (TE527)
Sqn Ldr Hugo Hrbácek
312 (Czechoslovak) Squadron, RAF
Praha-Ruzyne, Czechoslovakia, August 1945.

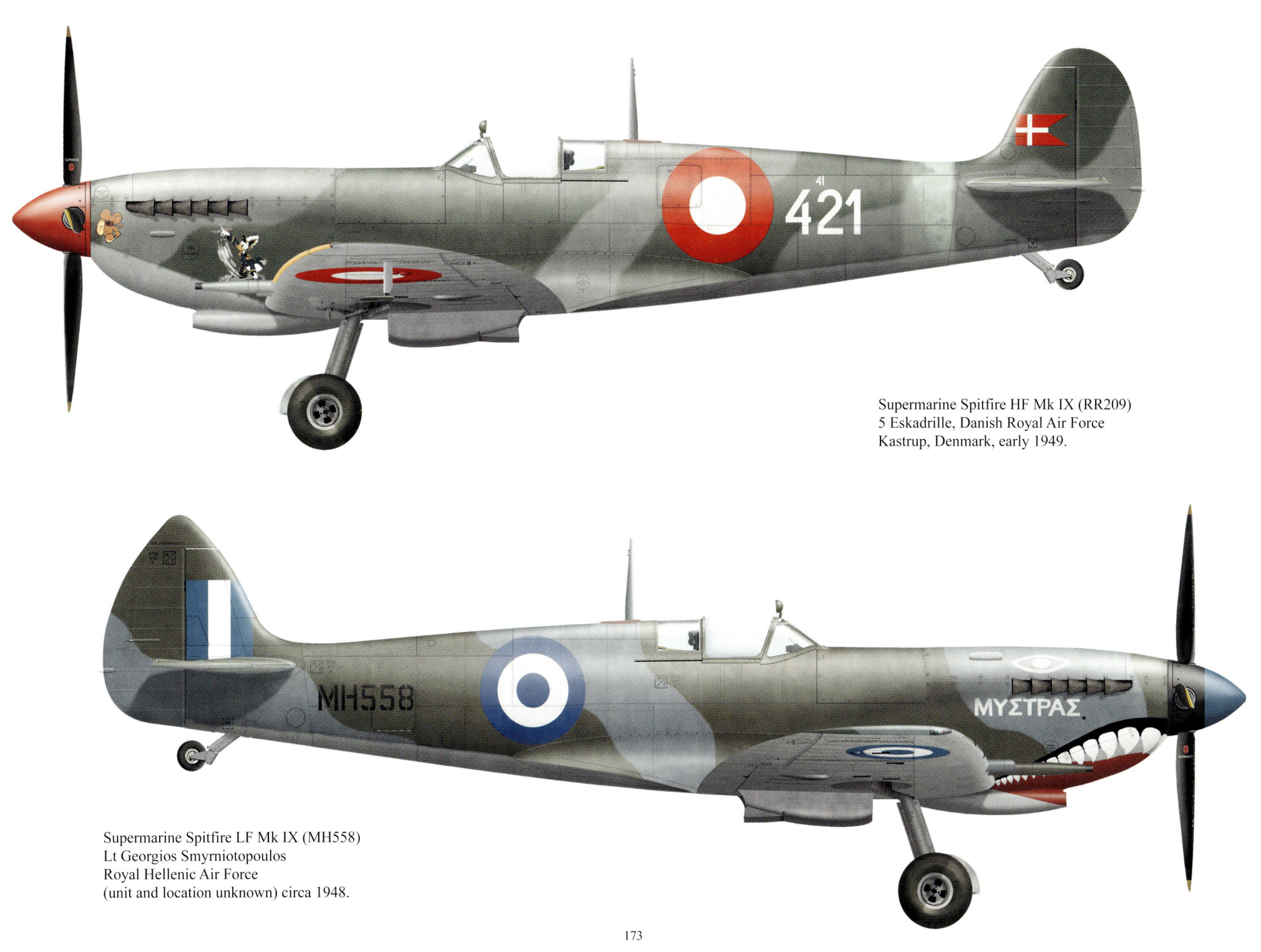

Supermarine Spitfire HF Mk IX (RR209)
5 Eskadrille, Danish Royal Air Force
Kastrup, Denmark, early 1949.

Supermarine Spitfire LF Mk IX (MH558)
Lt Georgios Smyrniotopoulos
Royal Hellenic Air Force
(unit and location unknown) circa 1948.

Supermarine Spitfire LF Mk IX (MJ893)
322e Jachtvliegtuig Afdeling, Royal Netherlands AF
Kalibanteng, Dutch East Indies, September 1948.

Supermarine Spitfire LF Mk IXe (MK791)
2eme Escadrille (SPA81), GC 1/4 'Dauphine', Armée de l'Air
Tan-Son-Nhut, Indochina, late 1948.
Note American 250lb bomb.

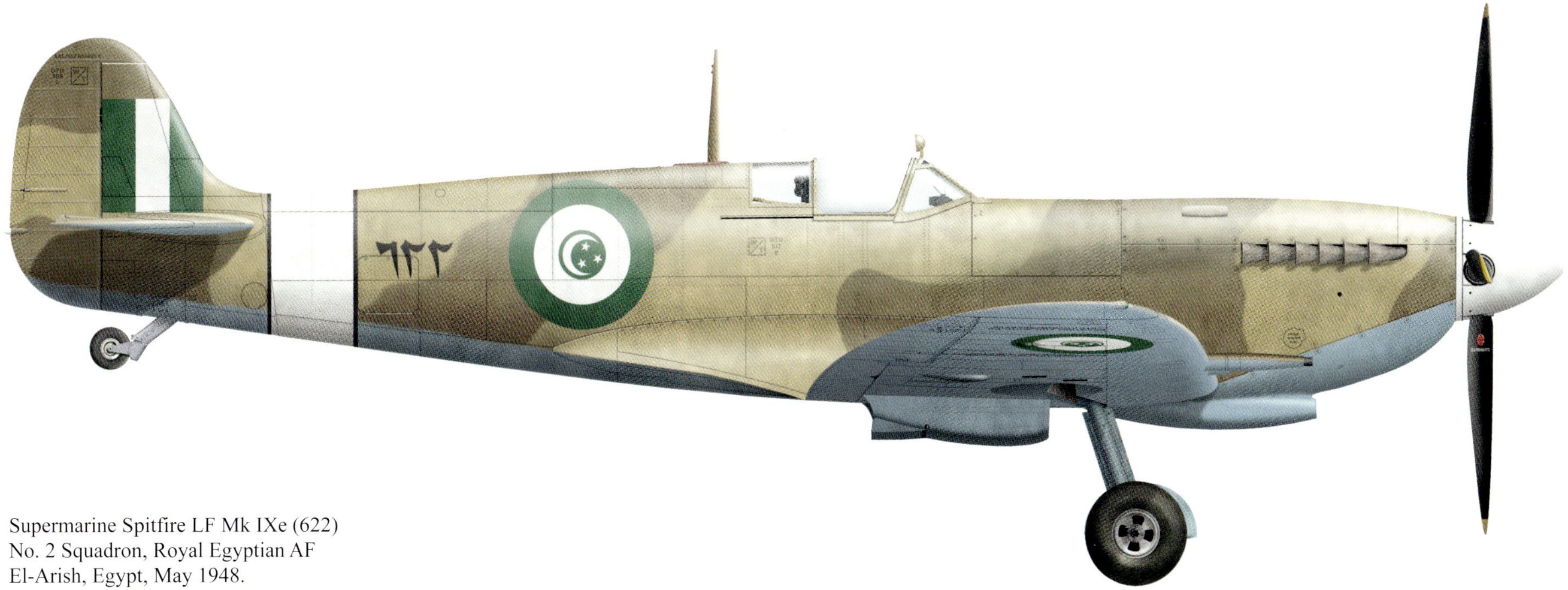

Supermarine Spitfire LF Mk IXe (622)
No. 2 Squadron, Royal Egyptian AF
El-Arish, Egypt, May 1948.

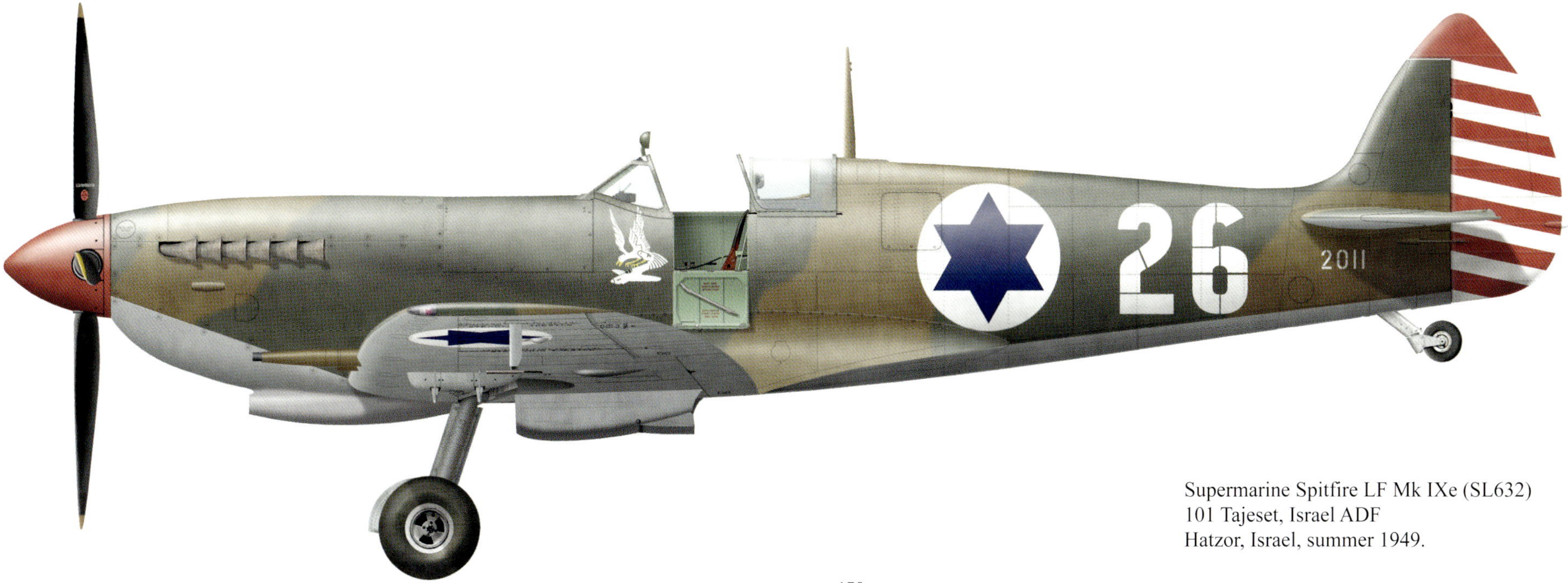

Supermarine Spitfire LF Mk IXe (SL632)
101 Tajeset, Israel ADF
Hatzor, Israel, summer 1949.

Supermarine Spitfire Mk IX (serial unknown)
92ª Squadriglia, 5º Stormo, Aeronautica Militare Italiana
Bergamo, Italy, late 1949.

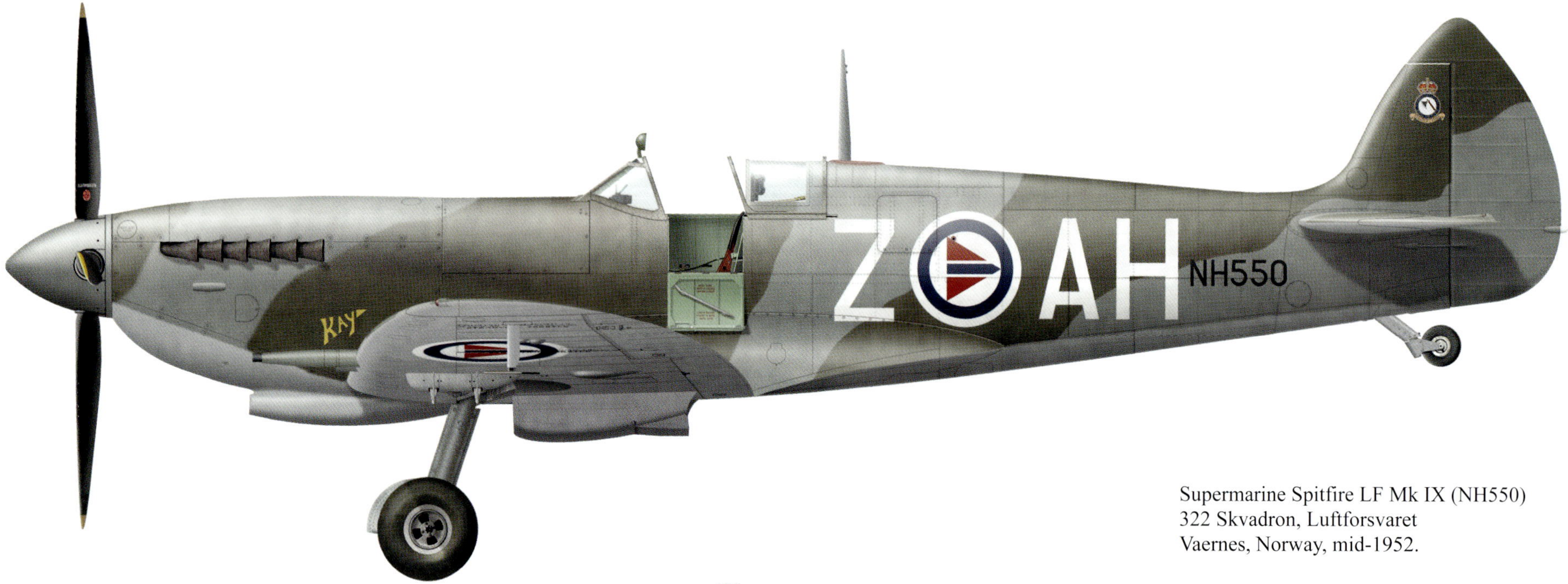

Supermarine Spitfire LF Mk IX (NH550)
322 Skvadron, Luftforsvaret
Vaernes, Norway, mid-1952.

Spitfire Mk XVI

Supermarine Spitfire F Mk XVIe (RR227)
Sqn Ldr Otto Smik
127 Squadron, RAF
Grimbengen, Belgium, November 1944.

Supermarine Spitfire LF Mk XVIe (SM311)
Flt Lt R. D. 'Dagwood' Philips
416 (Lynx) Squadron, RCAF
Evere, Belgium, January 1945.

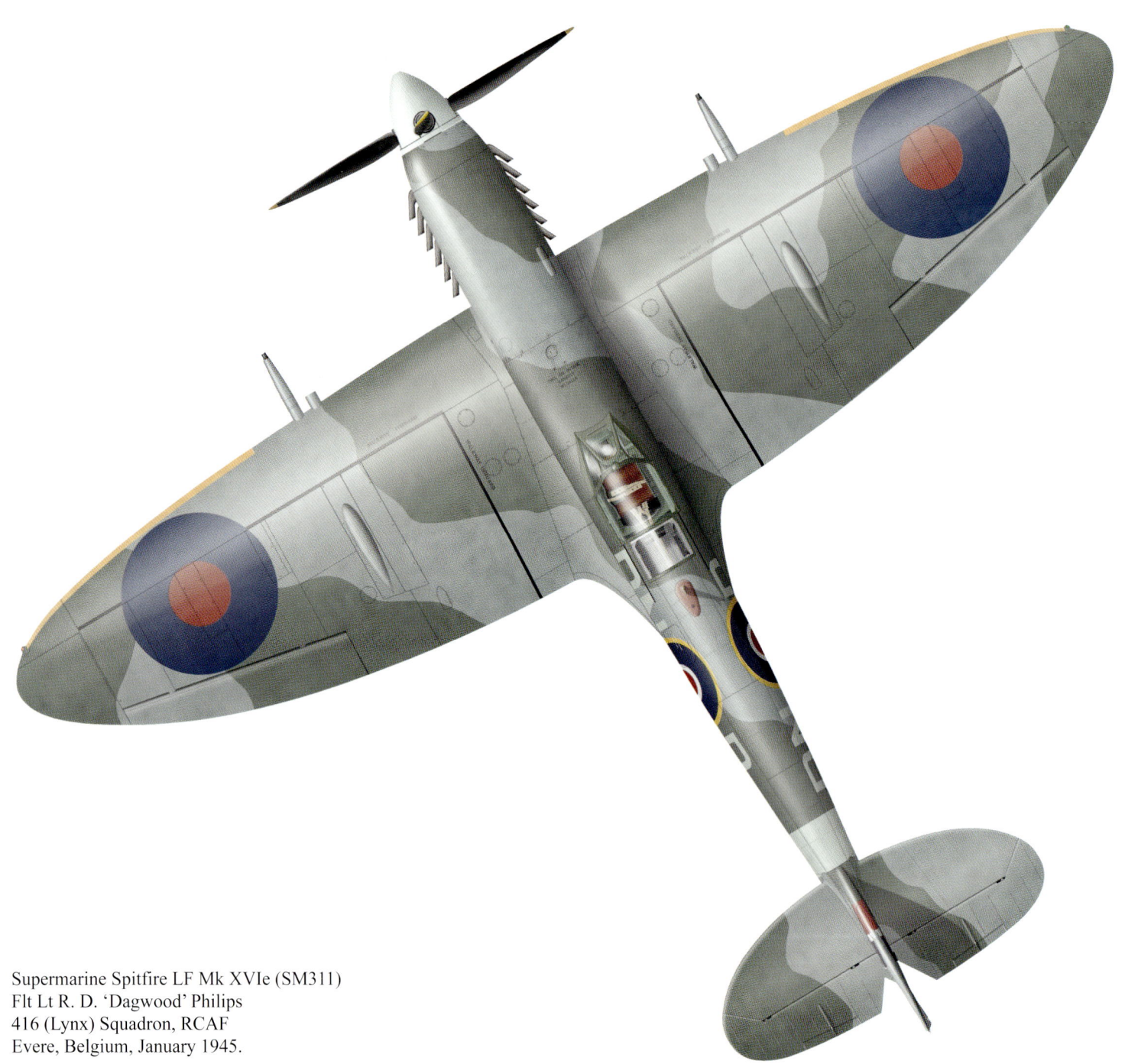

Supermarine Spitfire LF Mk XVIe (SM311)
Flt Lt R. D. 'Dagwood' Philips
416 (Lynx) Squadron, RCAF
Evere, Belgium, January 1945.

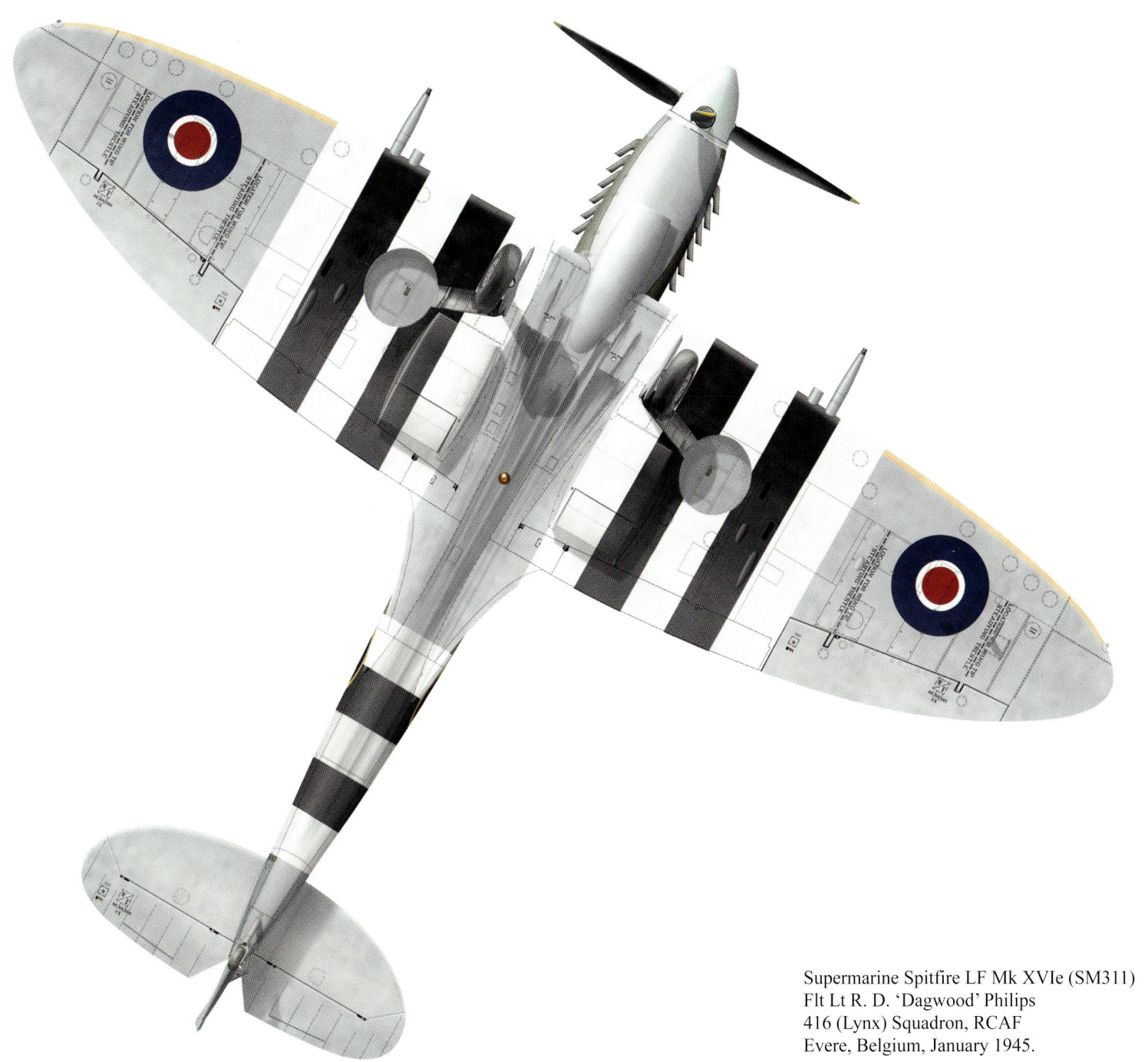

Supermarine Spitfire LF Mk XVIe (SM311)
Flt Lt R. D. 'Dagwood' Philips
416 (Lynx) Squadron, RCAF
Evere, Belgium, January 1945.

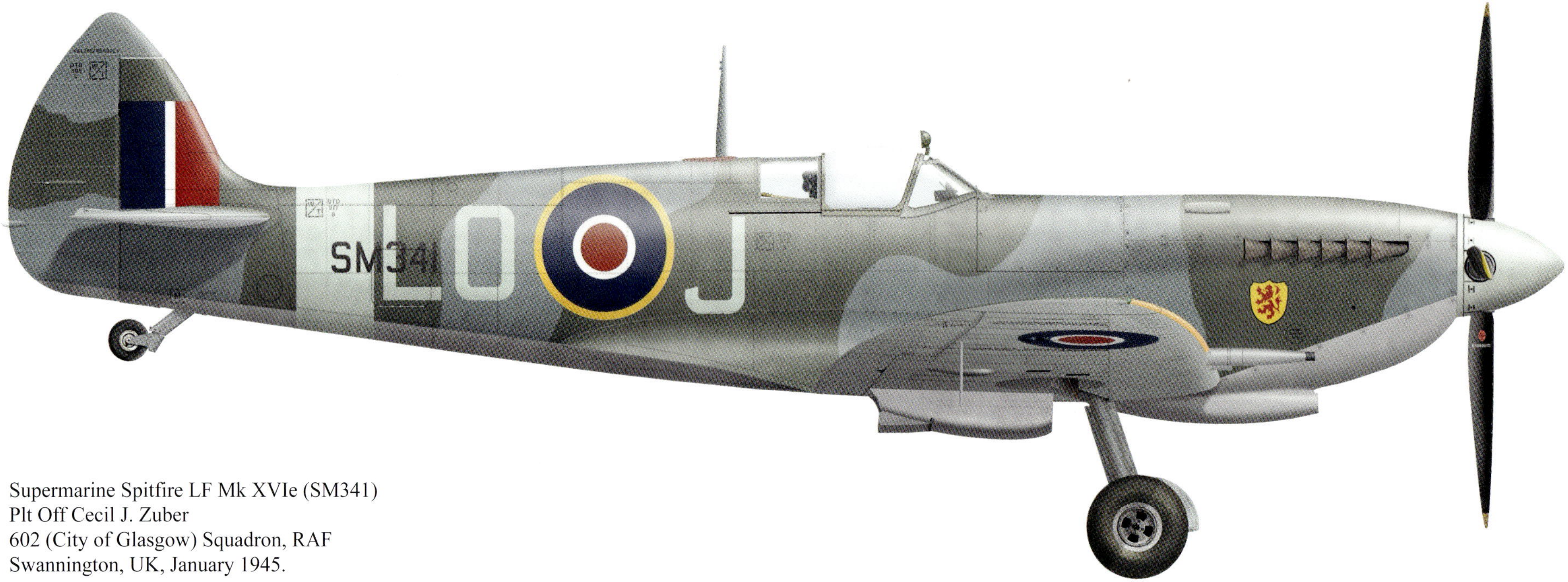

Supermarine Spitfire LF Mk XVIe (SM341)
Plt Off Cecil J. Zuber
602 (City of Glasgow) Squadron, RAF
Swannington, UK, January 1945.

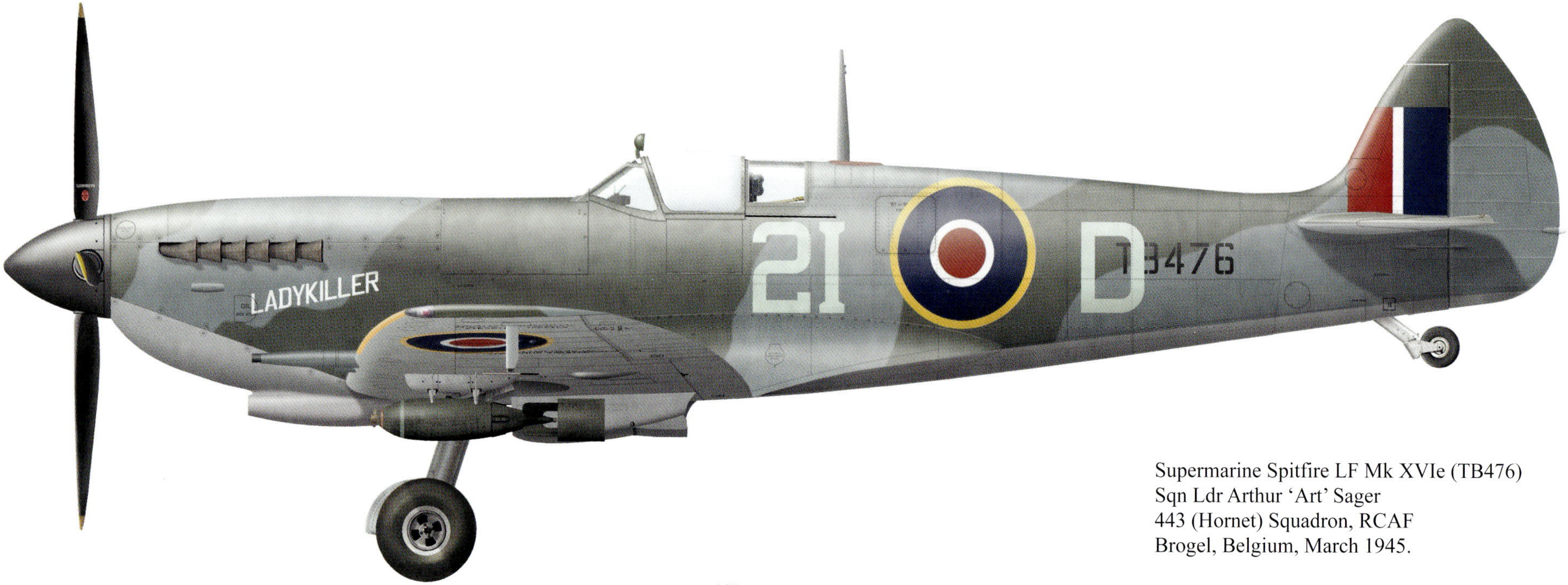

Supermarine Spitfire LF Mk XVIe (TB476)
Sqn Ldr Arthur 'Art' Sager
443 (Hornet) Squadron, RCAF
Brogel, Belgium, March 1945.

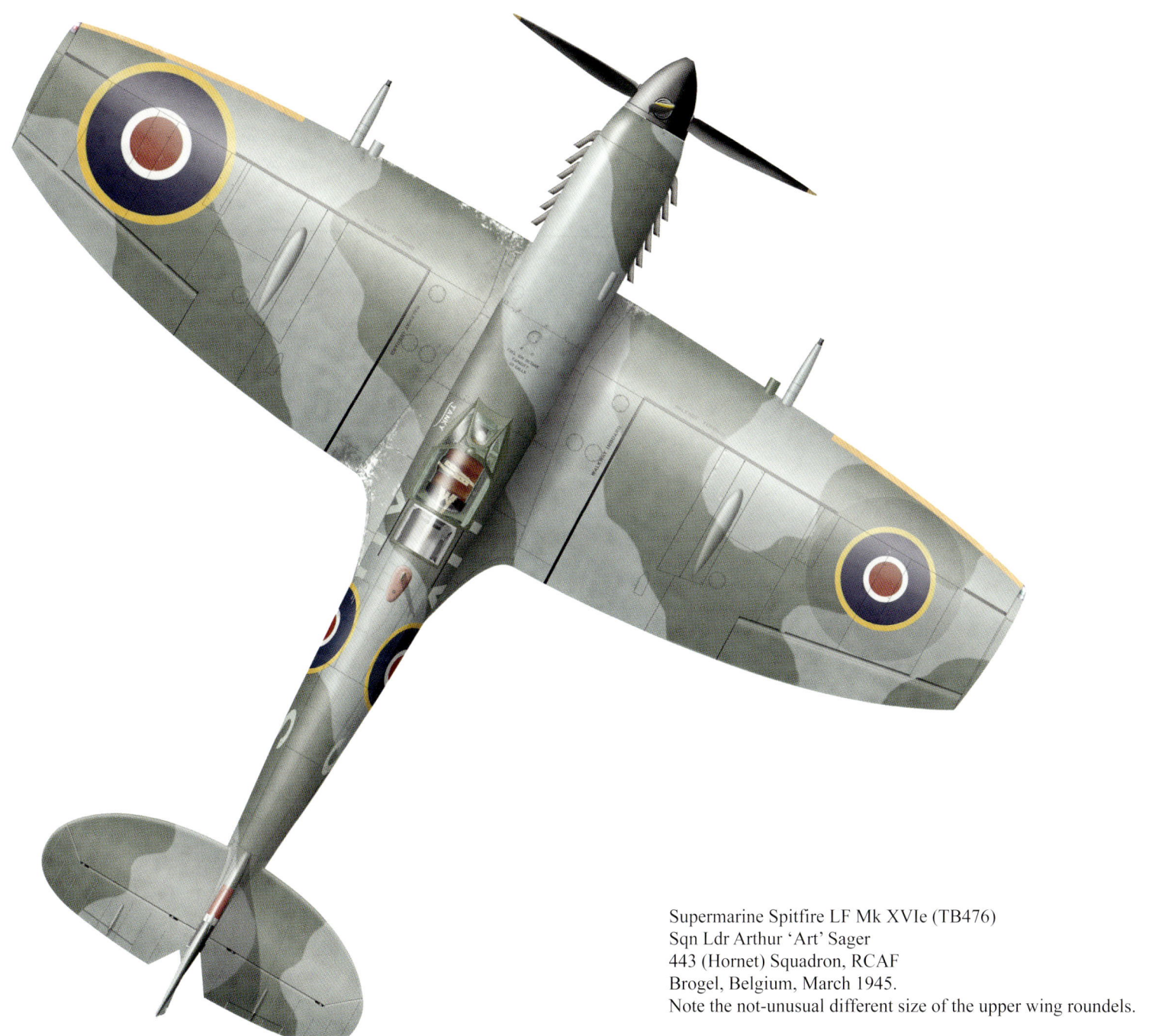

Supermarine Spitfire LF Mk XVIe (TB476)
Sqn Ldr Arthur 'Art' Sager
443 (Hornet) Squadron, RCAF
Brogel, Belgium, March 1945.
Note the not-unusual different size of the upper wing roundels.

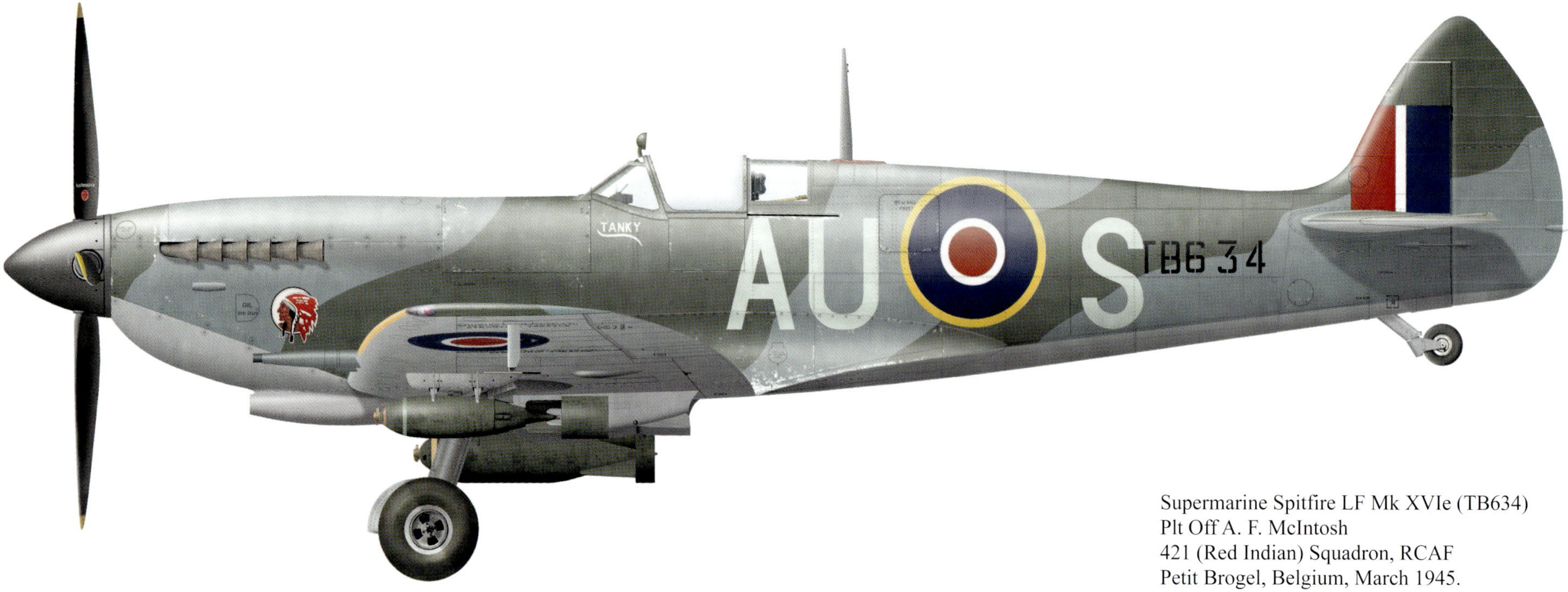

Supermarine Spitfire LF Mk XVIe (TB634)
Plt Off A. F. McIntosh
421 (Red Indian) Squadron, RCAF
Petit Brogel, Belgium, March 1945.

Supermarine Spitfire LF Mk. XVIe (TB300)
Gp Capt Stan Turner
127 Wing, RAF
Evere, Belgium, April 1945.

Supermarine Spitfire LF Mk XVIe (TB752)
Sqn Ldr Henry 'Hank' Zary
403 (City of Calgary) Squadron, RCAF
Diepholz, Germany, April 1945.

Supermarine Spitfire LF Mk XVIe (TB886)
Flt Lt Bill Harper
421 (Red Indian) Squadron, RCAF
Reinsehlen, Germany, May 1945.

Supermarine Spitfire LF Mk XVIe (TD338)
345 (Free French) Squadron, RAF
Drope, Germany, May 1945.

Supermarine Spitfire LF Mk XVIe (TB702)
Capt P. G. J. Albertin
340 (Free French) Squadron, RAF
Lingen/Drope, Germany, May 1945.

Supermarine Spitfire LF Mk XVIe (TD240)
Sqn Ldr Boleslaw Kaczmarek
302 (Polish) Squadron, RAF
Varrelsbuch, Germany, summer 1945.

Supermarine Spitfire LF Mk XVIe (TB900)
Sqn Ldr Raymond Lallemand
349 (Belgian) Squadron, RAF
Wunstorf, Germany, summer 1945.

Supermarine Spitfire LF Mk XVIe (TB675)
WO M. Lind
485 Squadron, RNZAF
Fassberg, Germany, summer 1945.

Supermarine Spitfire LF Mk XVIe (TD341)
443 (Hornet) Squadron, RCAF
Ueterse, Germany, August 1945.

Supermarine Spitfire LF Mk XVIe (TD317)
Sqn Ldr Karol Pniak
308 (Polish) Squadron, RAF
Ahlhorn, Germany, September 1945.

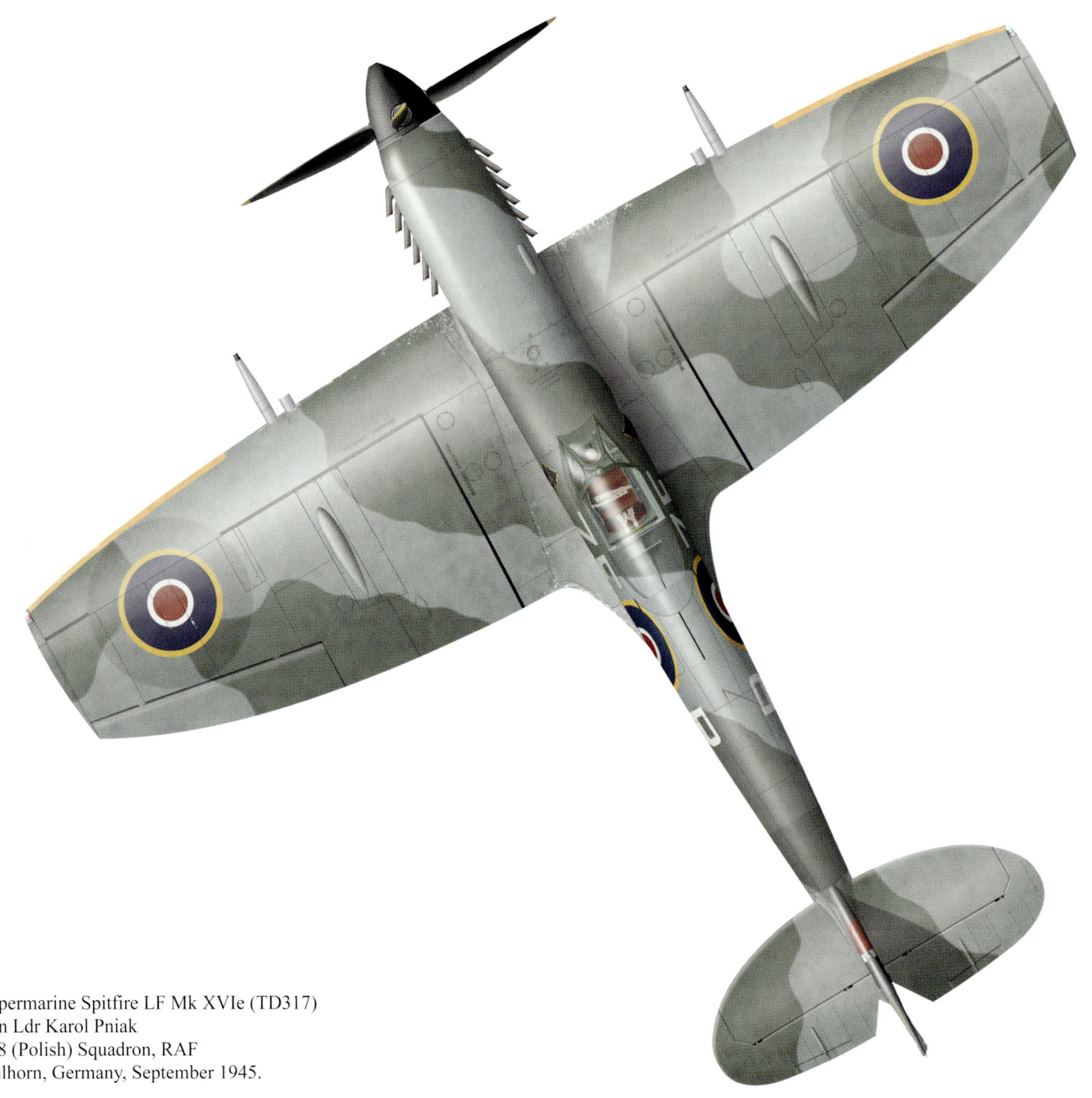

Supermarine Spitfire LF Mk XVIe (TD317)
Sqn Ldr Karol Pniak
308 (Polish) Squadron, RAF
Ahlhorn, Germany, September 1945.

Supermarine Spitfire LF Mk XVIe (TD317)
Sqn Ldr Karol Pniak
308 (Polish) Squadron, RAF
Ahlhorn, Germany, September 1945.

Supermarine Spitfire LF Mk XVIe (TD188)
350 (Belgian) Squadron, RAF
Fassberg, Germany, summer 1946.

Supermarine Spitfire LF Mk XVIe (TD237)
349e Escadrille, Aviation Militaire Belge
Beauvchain, Belgium, October 1946.

Supermarine Spitfire LF Mk XVIe (SL727)
601 (County of London) Squadron, RAF
North Weald, UK, late 1949.

Supermarine Spitfire LF Mk XVIe (SL727)
601 (County of London) Squadron, RAF
North Weald, UK, late 1949.

Supermarine Spitfire LF Mk XVIe (SL727)
601 (County of London) Squadron, RAF
North Weald, UK, late 1949.

Supermarine Spitfire LF Mk XVIe (SL549)
17 Squadron, RAF
Farnborough, UK, circa 1950.

Supermarine Spitfire LF Mk XVI (5621)
Air Operations School
Langebaanweg, South Africa, circa 1953.

PR Spitfires

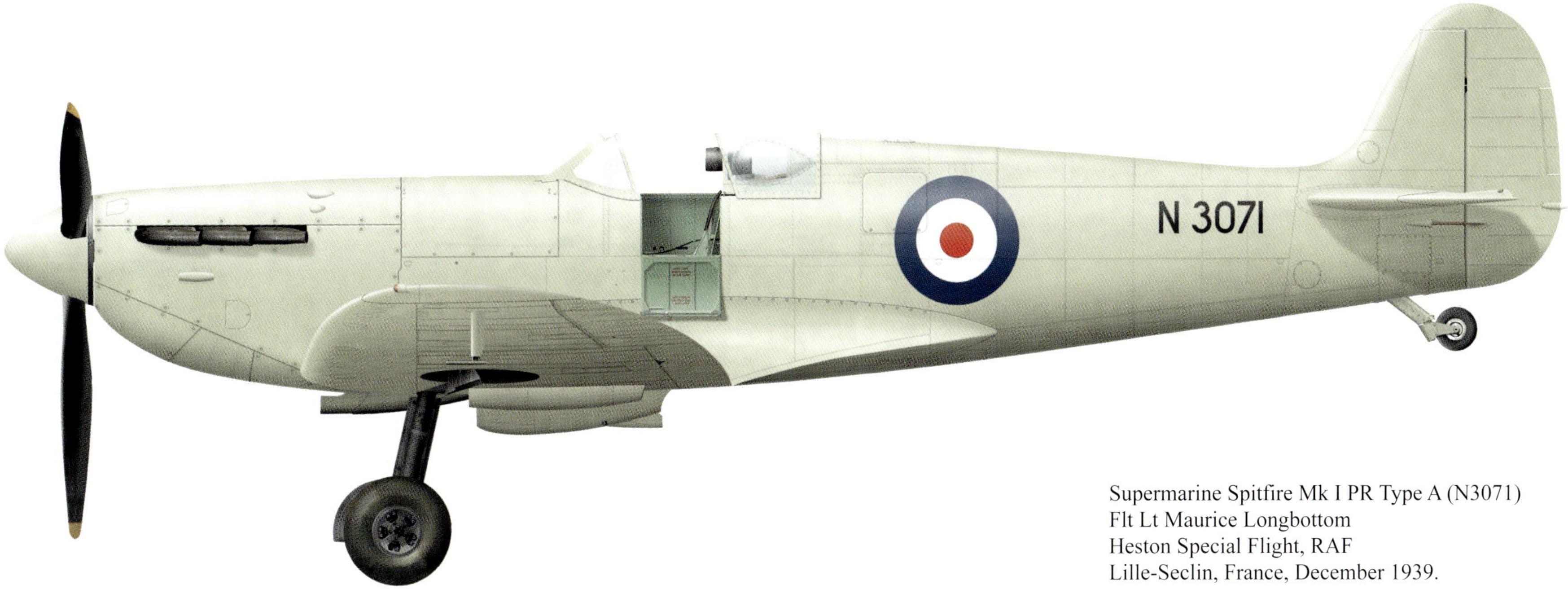

Supermarine Spitfire Mk I PR Type A (N3071)
Flt Lt Maurice Longbottom
Heston Special Flight, RAF
Lille-Seclin, France, December 1939.

Supermarine Spitfire Mk I PR Type D (P9551)
Fg Off Zbigiew Wisiekierski
1 Photographic Reconnaissance Unit, RAF
Benson, UK, December 1940.

Supermarine Spitfire Mk I Type C (R6903)
Fg Off G. Green
1 Photographic Reconnaissance Unit, RAF
Benson, UK, March 1941.

Supermarine Spitfire Mk I PR Type G (K9969)
1416 Flight, RAF
Hendon, UK, April 1941.

Supermarine Spitfire Mk I PR Type G (R7059)
Plt Off J. T. Morgan
1 Photographic Reconnaissance Unit
St Eval, UK, May 1941.

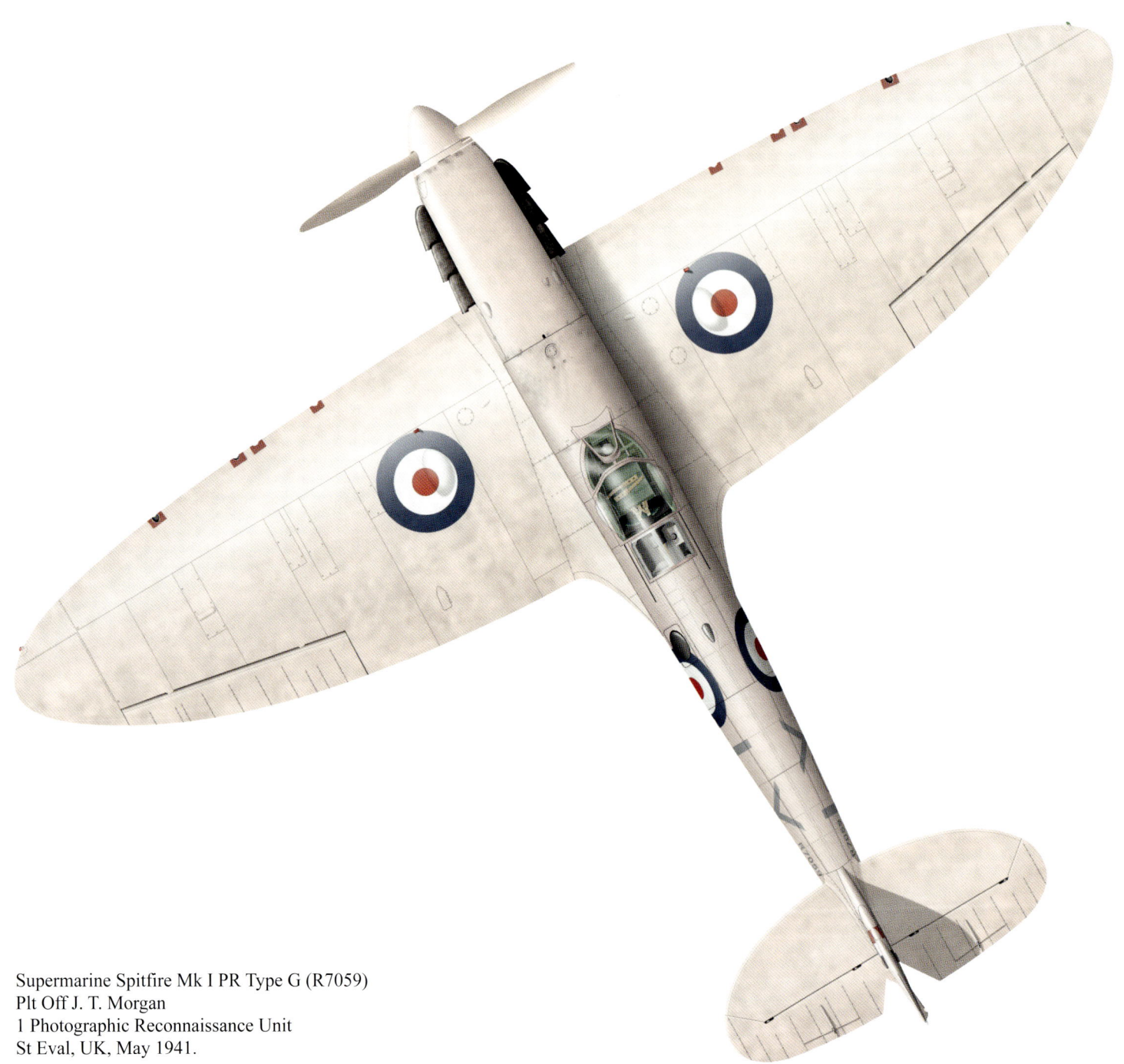

Supermarine Spitfire Mk I PR Type G (R7059)
Plt Off J. T. Morgan
1 Photographic Reconnaissance Unit
St Eval, UK, May 1941.

Supermarine Spitfire Mk I PR Type G (R7059)
Plt Off J. T. Morgan
1 Photographic Reconnaissance Unit
St Eval, UK, May 1941.

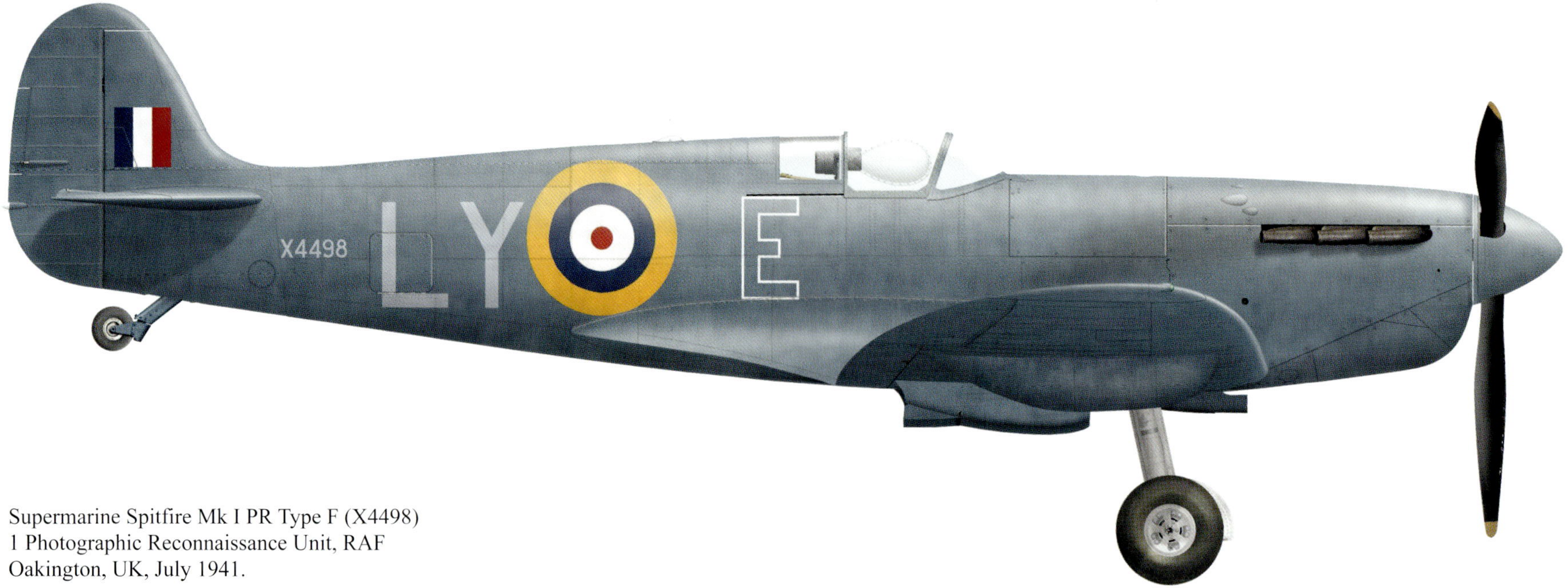

Supermarine Spitfire Mk I PR Type F (X4498)
1 Photographic Reconnaissance Unit, RAF
Oakington, UK, July 1941.

Supermarine Spitfire Mk I PR Type G (serial unknown)
140 Squadron, RAF
Benson, UK, September 1941.

Supermarine Spitfire Mk I PR Type G (R7142)
Plt Off C. A. P. Christie
140 Squadron, RAF
Benson, UK, December 1941.

Supermarine Spitfire PR Mk VII (X4784)
Plt Off C. B. Barber
140 Squadron, RAF
Benson, UK, April 1942.

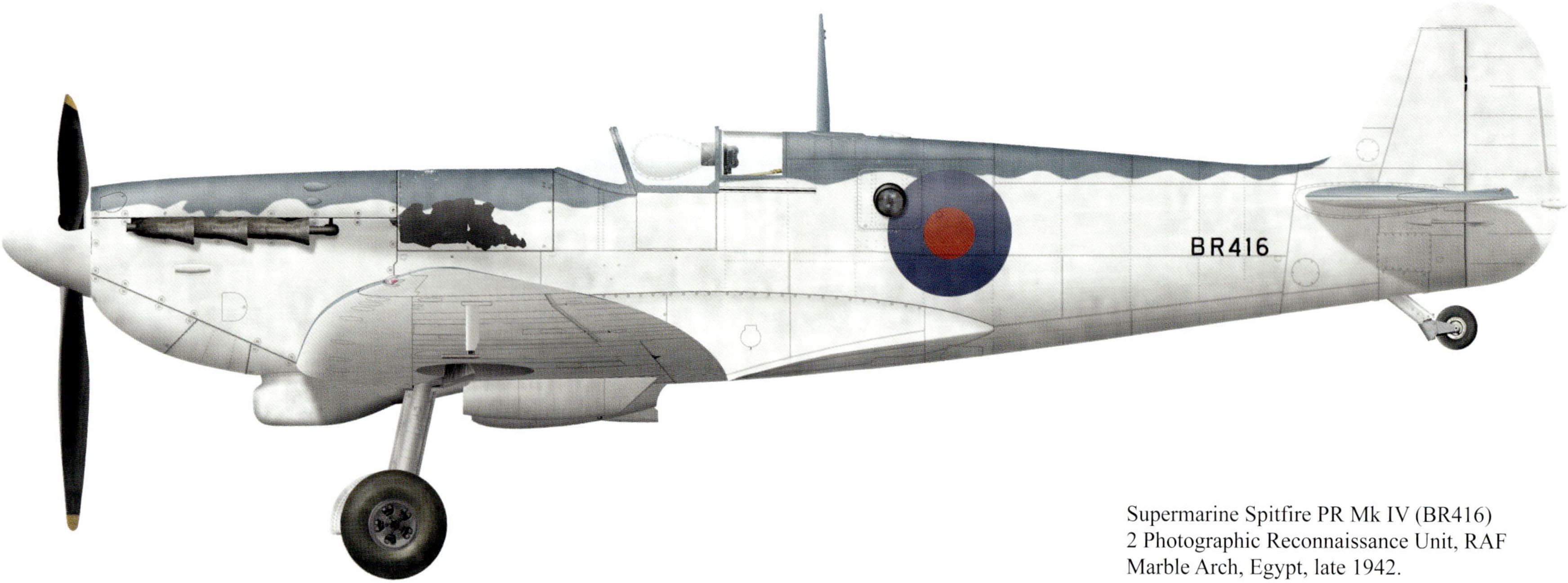

Supermarine Spitfire PR Mk IV (BR416)
2 Photographic Reconnaissance Unit, RAF
Marble Arch, Egypt, late 1942.

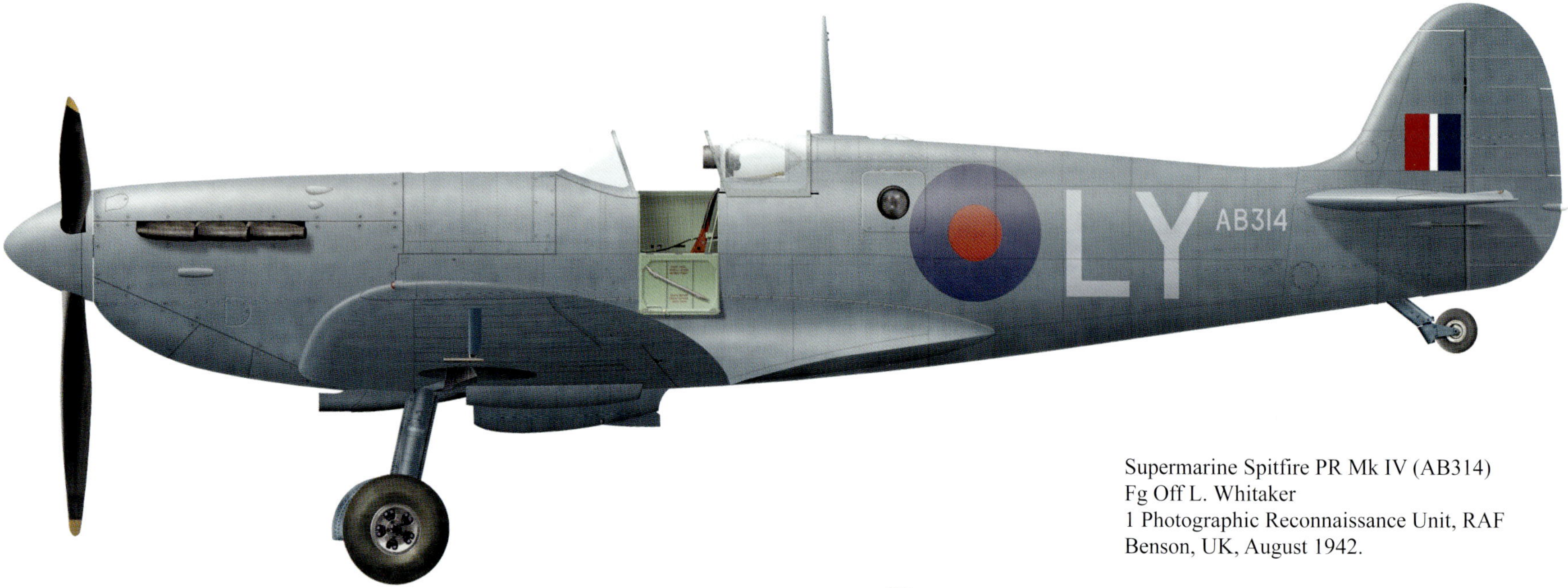

Supermarine Spitfire PR Mk IV (AB314)
Fg Off L. Whitaker
1 Photographic Reconnaissance Unit, RAF
Benson, UK, August 1942.

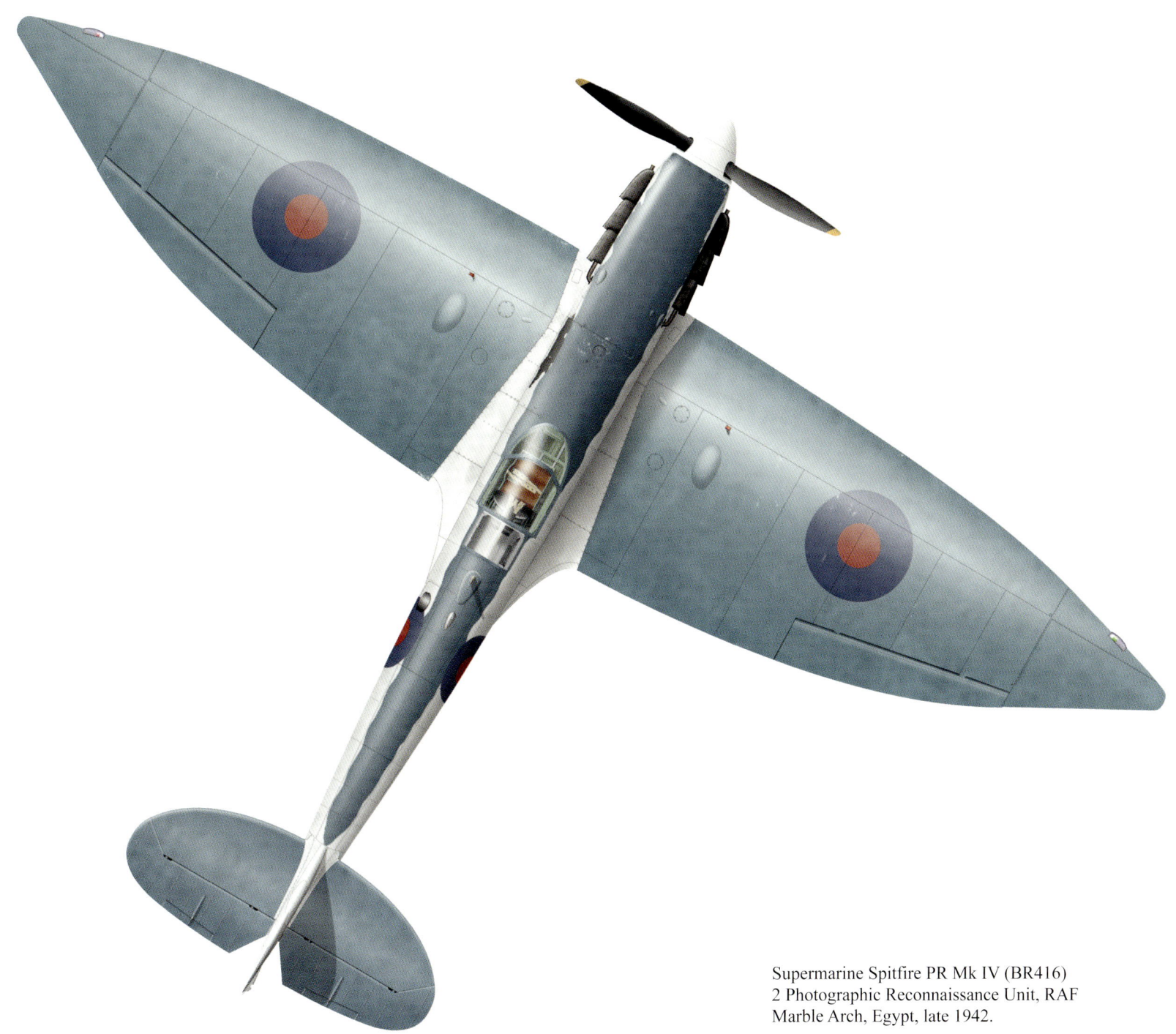

Supermarine Spitfire PR Mk IV (BR416)
2 Photographic Reconnaissance Unit, RAF
Marble Arch, Egypt, late 1942.

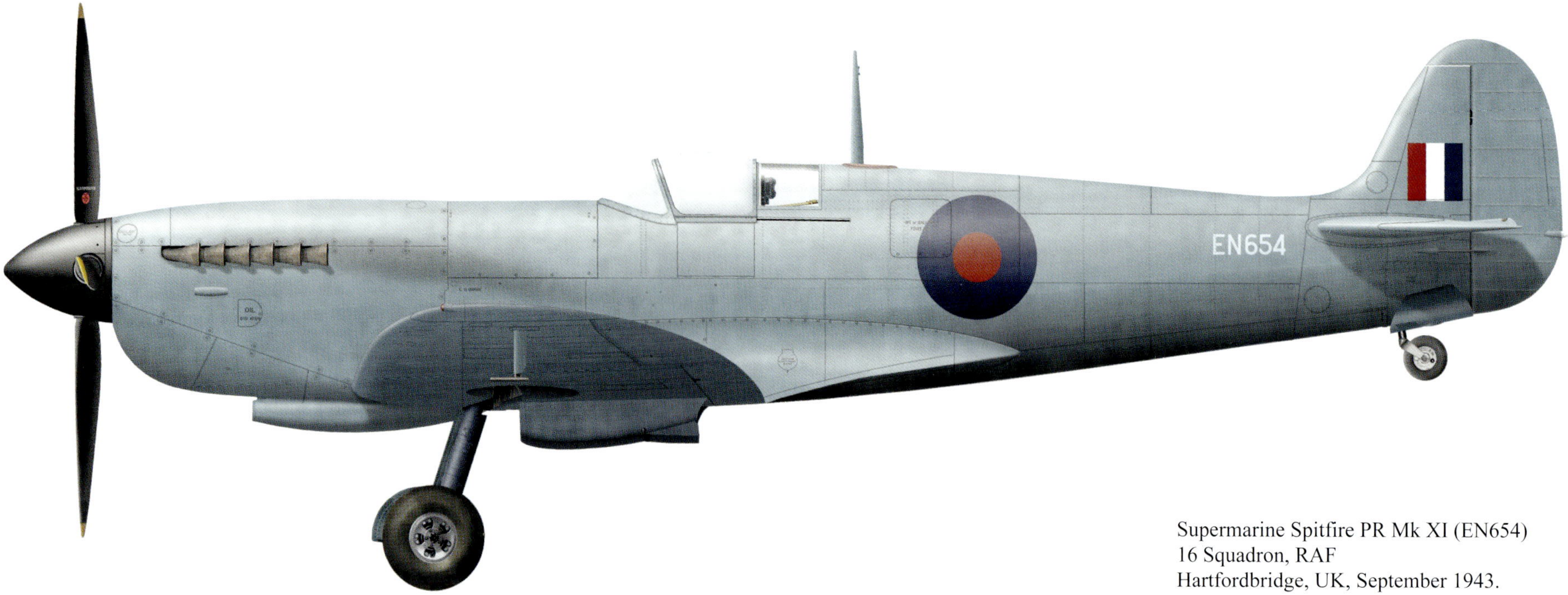

Supermarine Spitfire PR Mk XI (EN654)
16 Squadron, RAF
Hartfordbridge, UK, September 1943.

Supermarine Spitfire PR Mk IV (BP880)
Sgt Ron Monkman
681 Squadron, RAF
Chandina, Burma, February 1944.

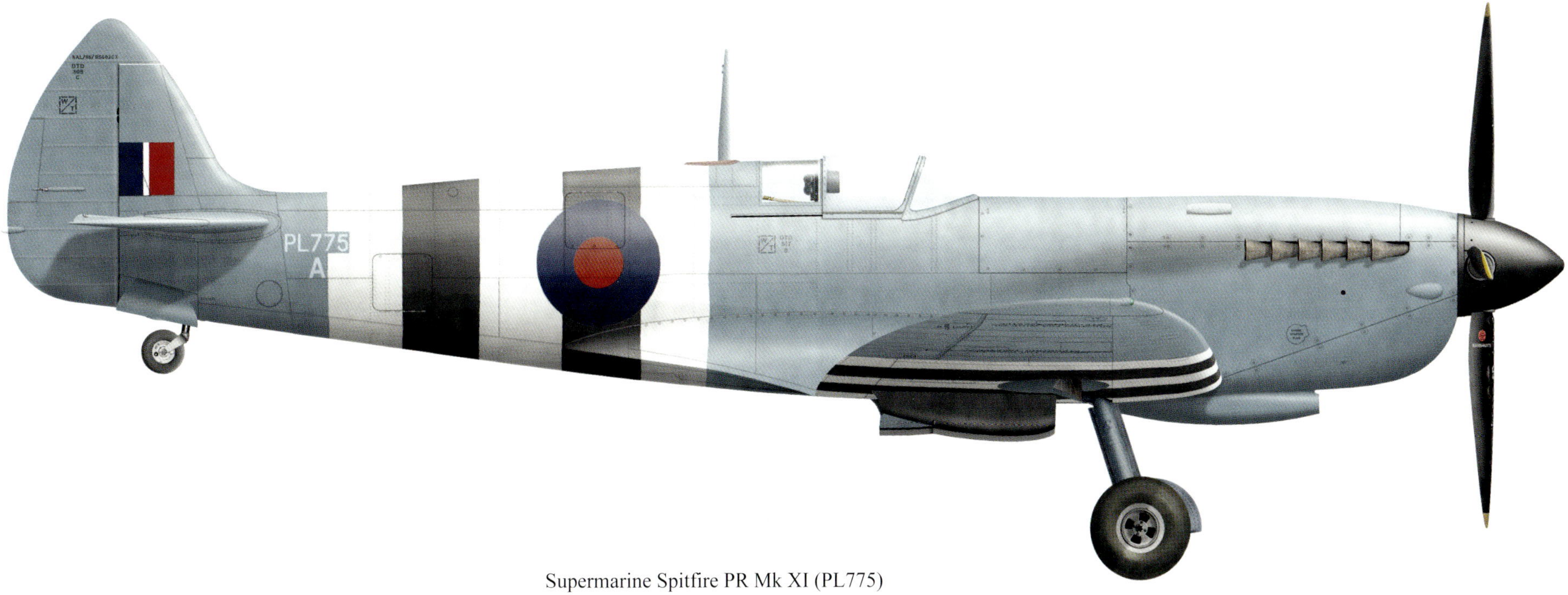

Supermarine Spitfire PR Mk XI (PL775)
541 Squadron, RAF
Benson, UK, June 1944.

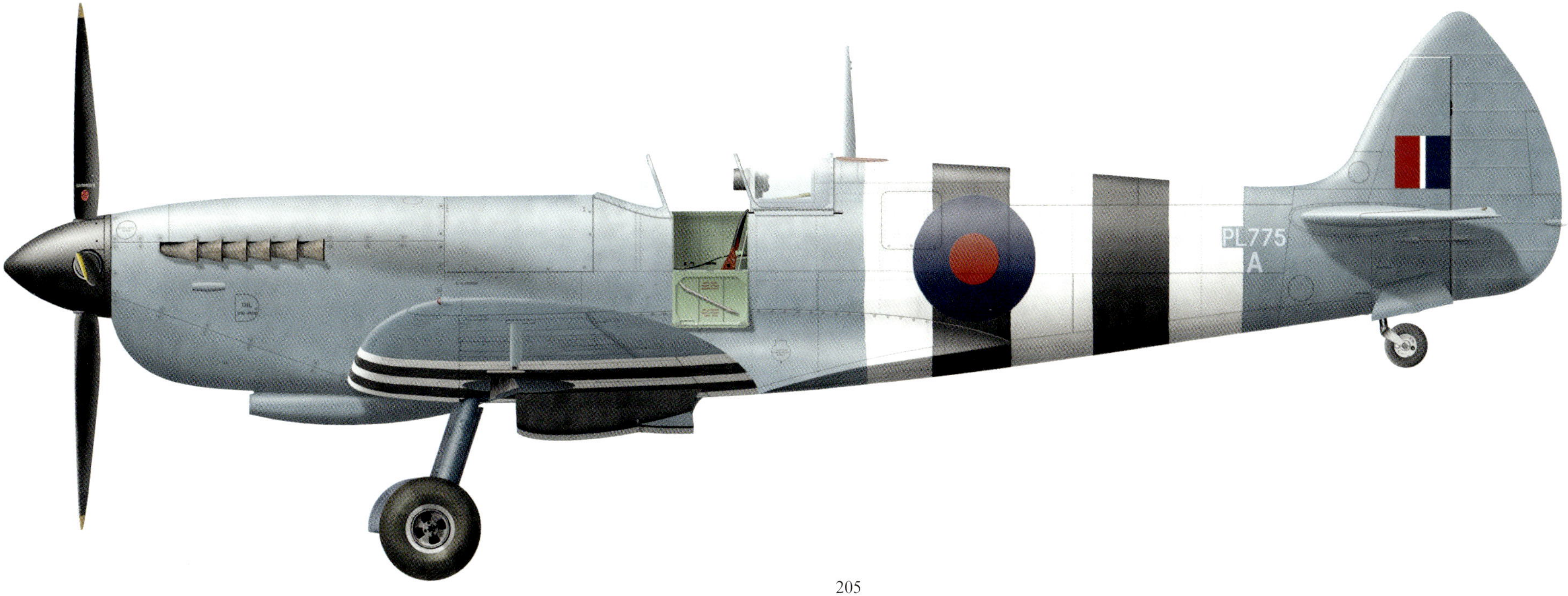

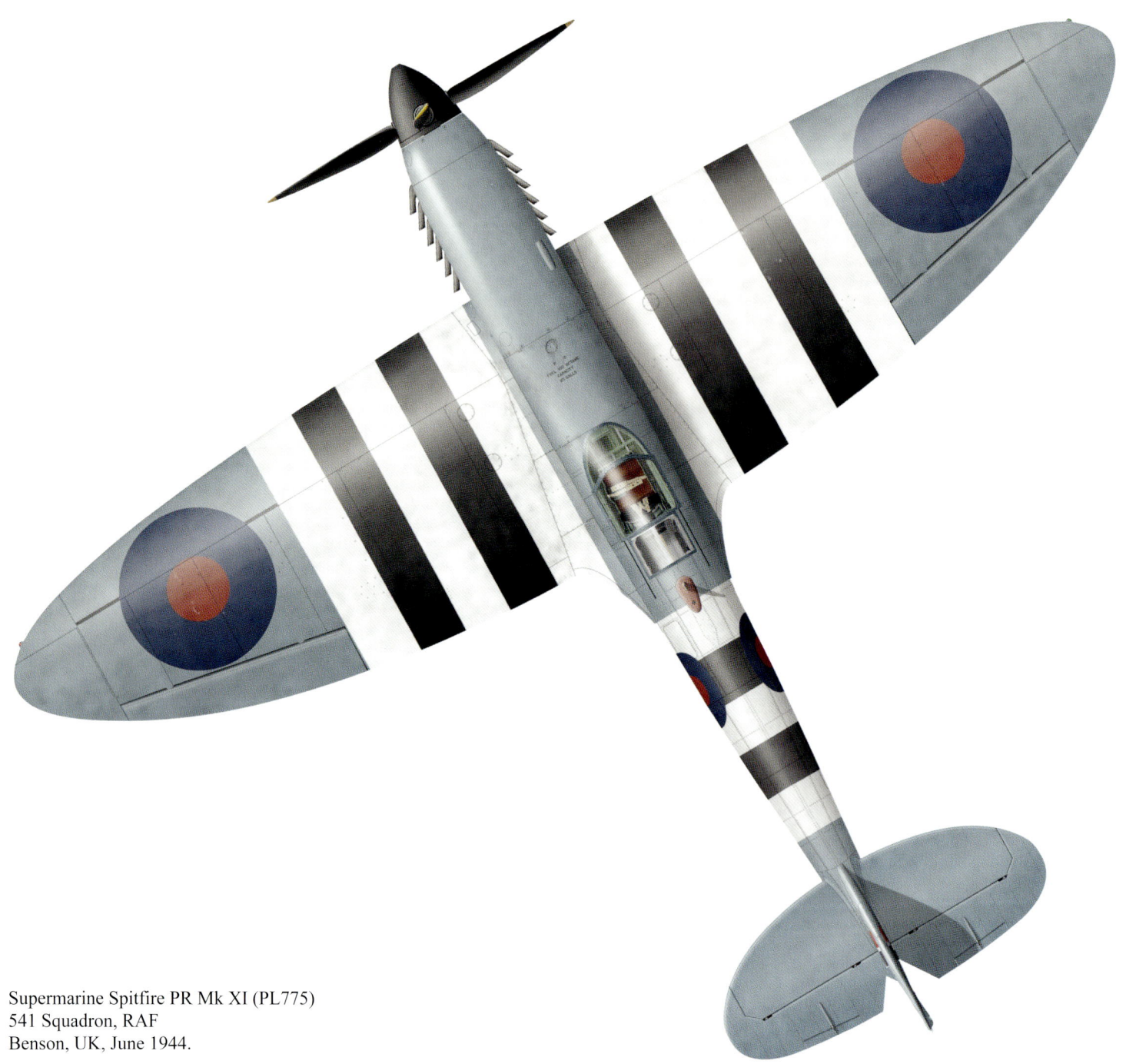

Supermarine Spitfire PR Mk XI (PL775)
541 Squadron, RAF
Benson, UK, June 1944.

Supermarine Spitfire PR Mk XI (PL775)
541 Squadron, RAF
Benson, UK, June 1944.

Supermarine Spitfire PR Mk X (SR396)
542 Squadron, RAF
Benson, UK, summer 1944.

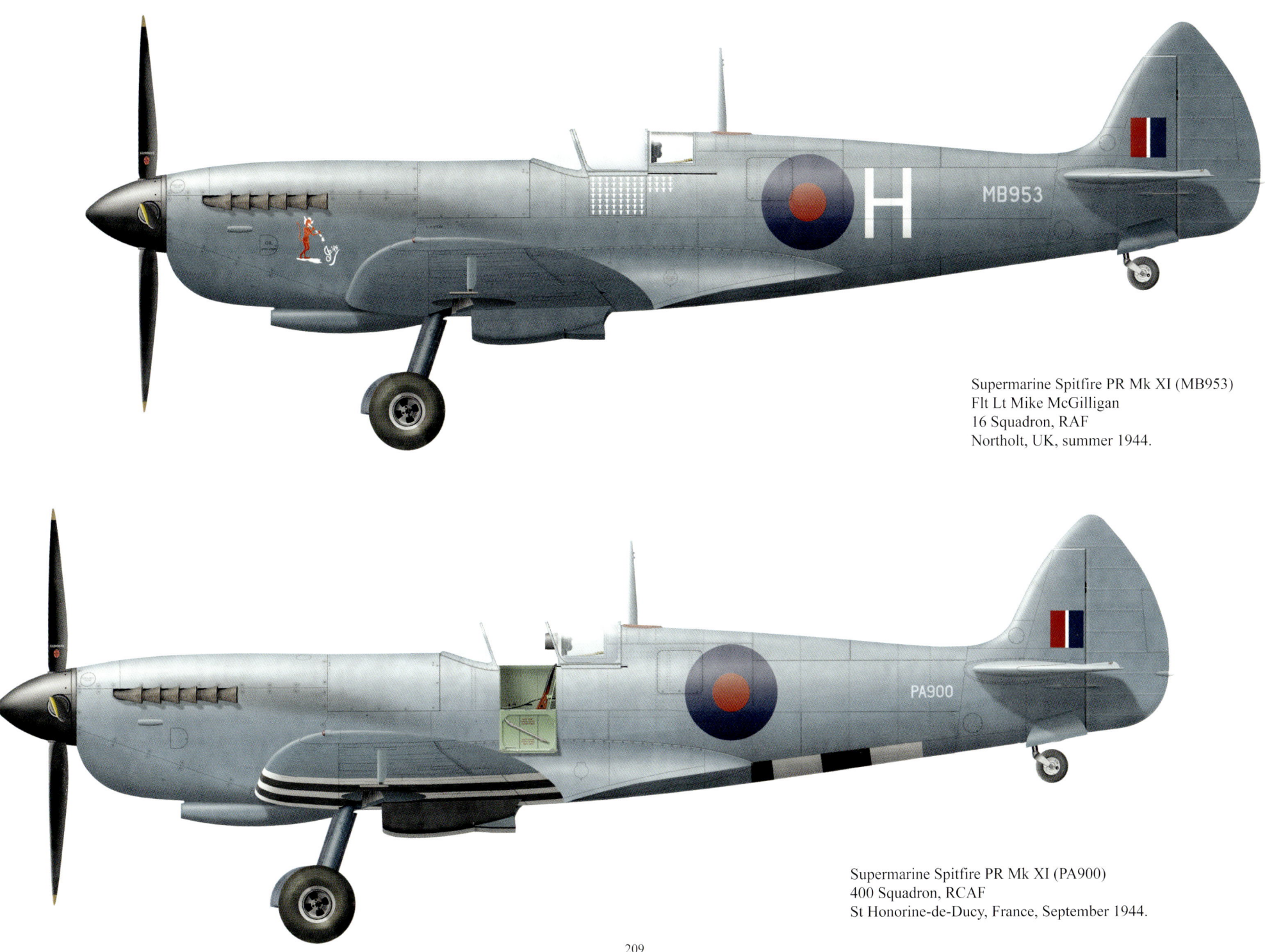

Supermarine Spitfire PR Mk XI (MB953)
Flt Lt Mike McGilligan
16 Squadron, RAF
Northolt, UK, summer 1944.

Supermarine Spitfire PR Mk XI (PA900)
400 Squadron, RCAF
St Honorine-de-Ducy, France, September 1944.

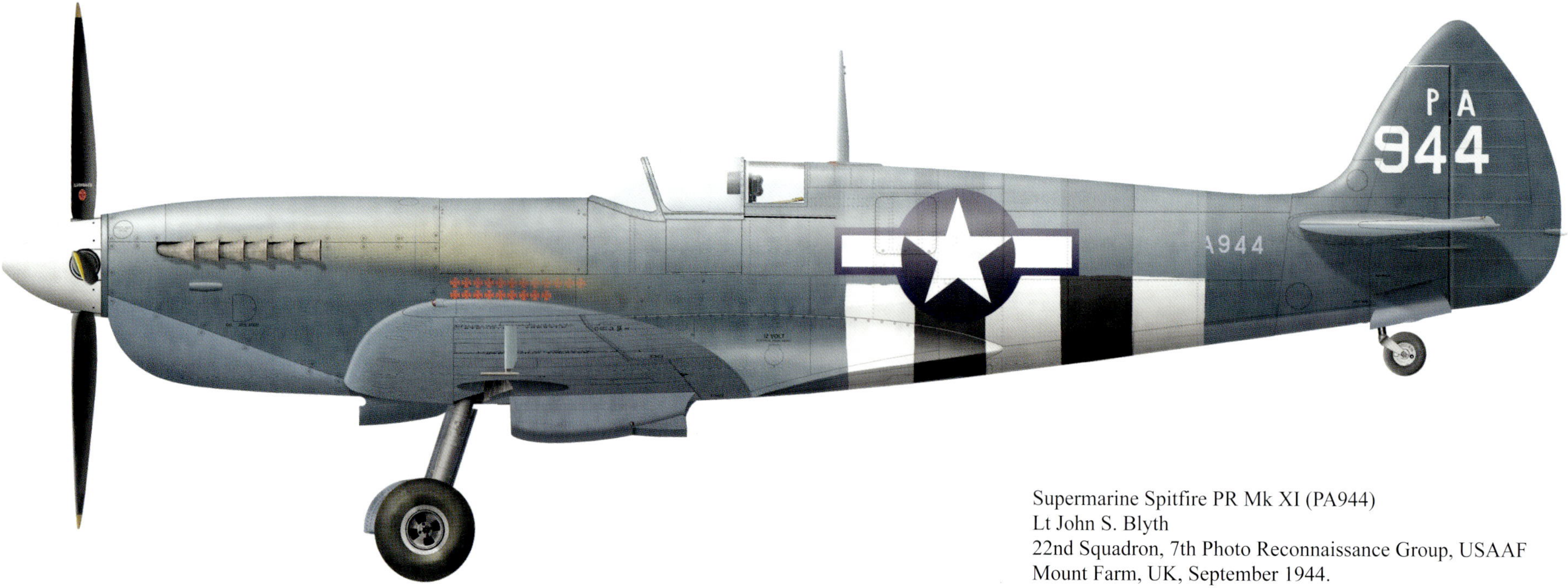

Supermarine Spitfire PR Mk XI (PA944)
Lt John S. Blyth
22nd Squadron, 7th Photo Reconnaissance Group, USAAF
Mount Farm, UK, September 1944.

Supermarine Spitfire PR Mk XI (PL863)
WO R. K. Brown
681 Squadron, RAF
Imphal, Burma, October 1944.

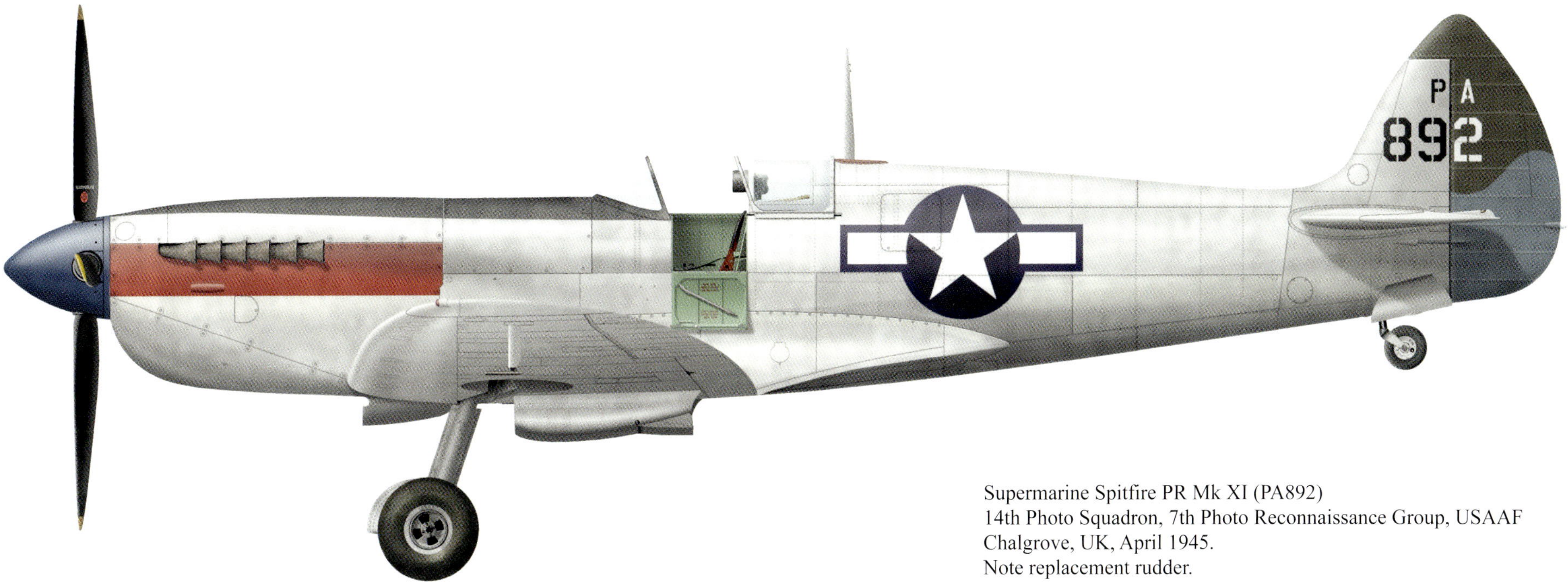

Supermarine Spitfire PR Mk XI (PA892)
14th Photo Squadron, 7th Photo Reconnaissance Group, USAAF
Chalgrove, UK, April 1945.
Note replacement rudder.

Supermarine Spitfire PR Mk XI (PL781)
681 Squadron, RAF
Kuala Lumpur, Malaya, late 1945.

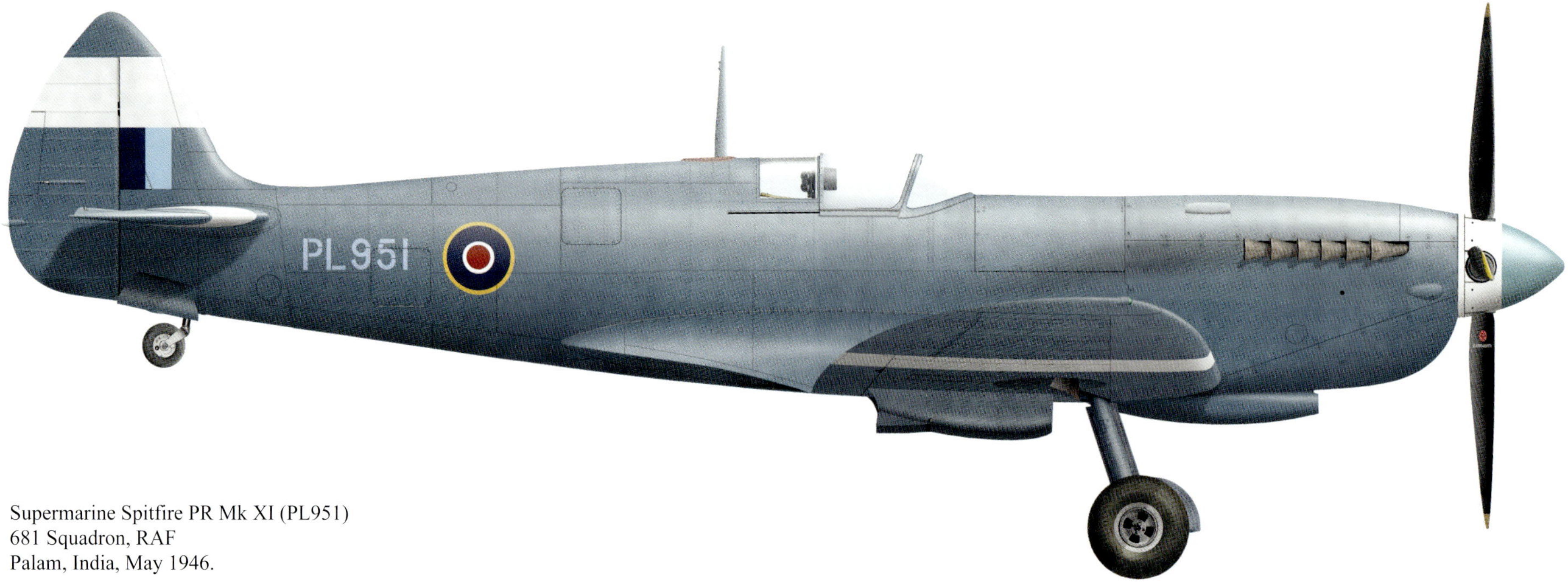

Supermarine Spitfire PR Mk XI (PL951)
681 Squadron, RAF
Palam, India, May 1946.

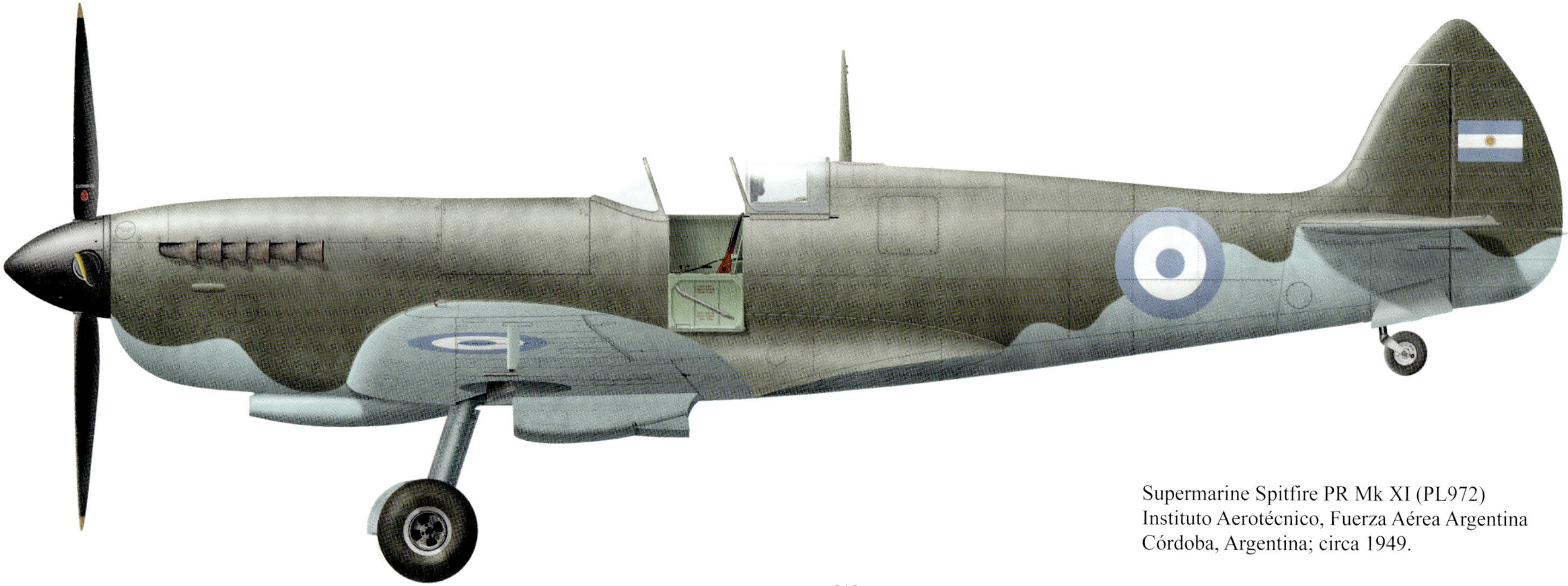

Supermarine Spitfire PR Mk XI (PL972)
Instituto Aerotécnico, Fuerza Aérea Argentina
Córdoba, Argentina; circa 1949.

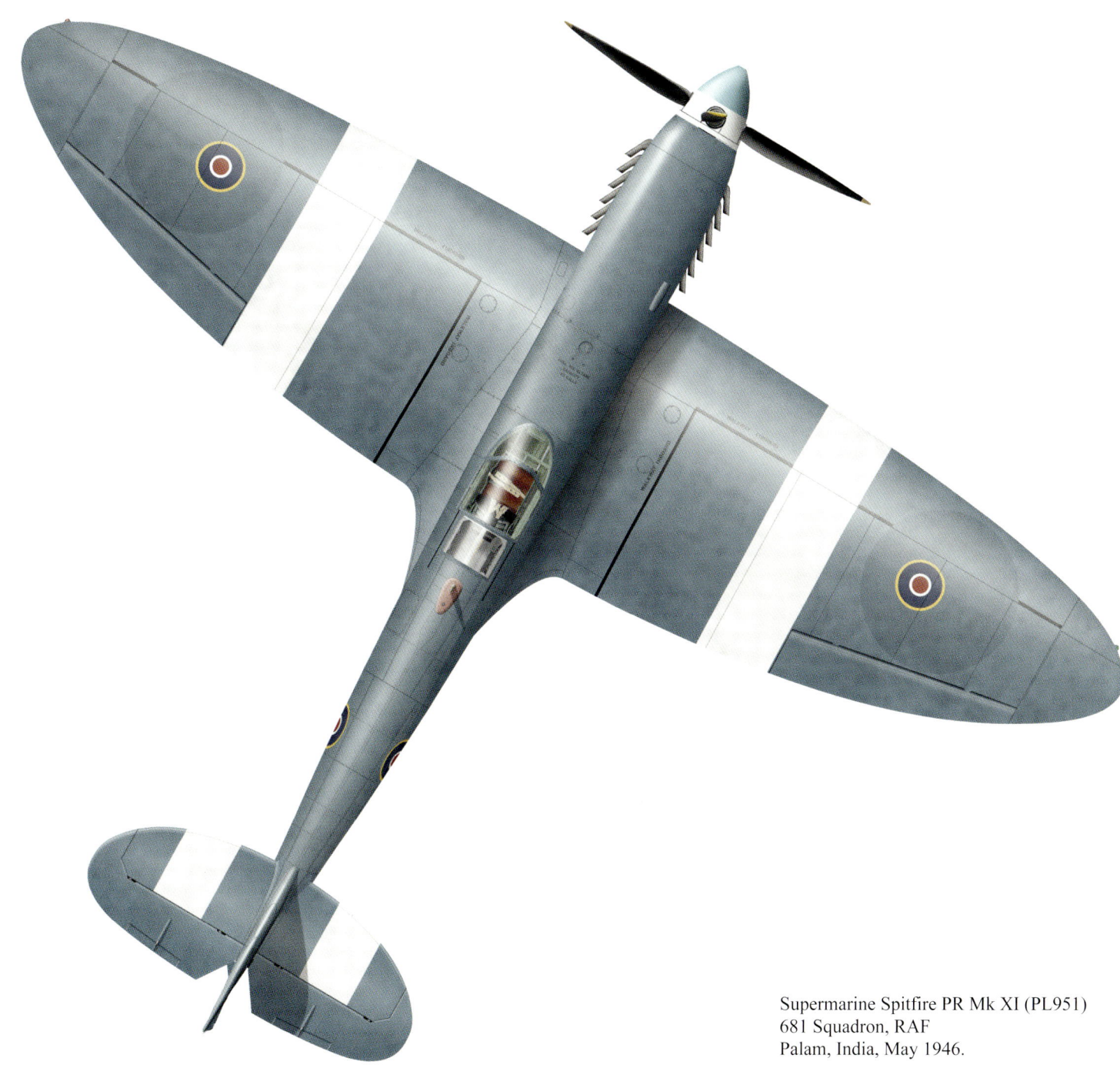

Supermarine Spitfire PR Mk XI (PL951)
681 Squadron, RAF
Palam, India, May 1946.

Supermarine Spitfire PR Mk XI (PL979)
Nr. 1 Fotorekognoseringsving, Luftforstvaret
Bardufoss, Norway, summer 1950.

Supermarine Spitfire PR Mk XI (42/451)
No. 722 Squadron, Danish Air Force
Vaerlose, Denmark, circa 1950.

Seafire

Supermarine Seafire Mk Ib (MB366)
801 Squadron, FAA
HMS *Furious*, Western Mediterranean, November 1942.
During Operation Torch, British aircraft wore American markings for political reasons.

Supermarine Seafire Mk IIc (MB156)
Sub Lt J. D. Buchanan
885 Squadron, FAA
HMS *Formidable*, Western Mediterranean, November 1942.

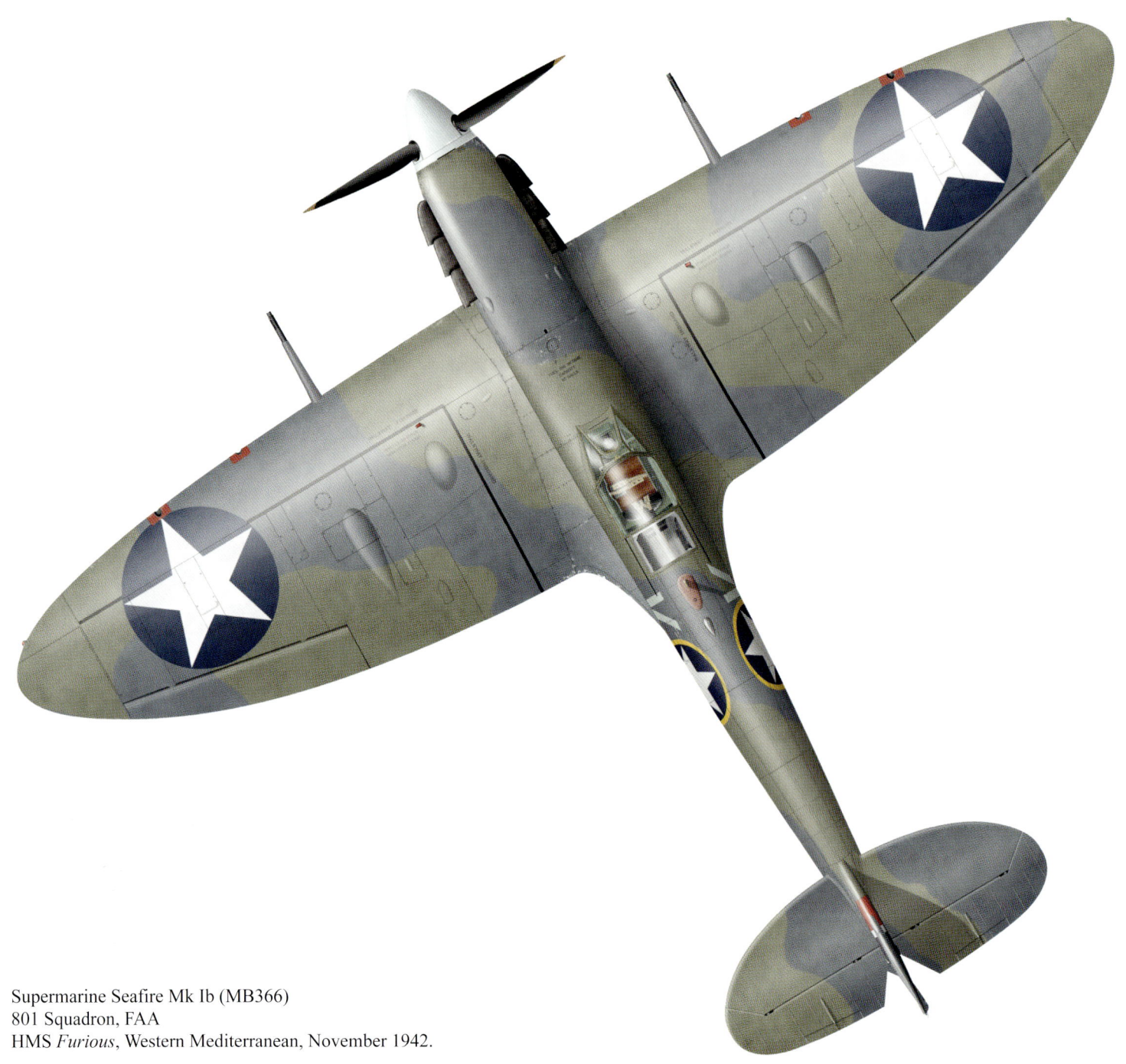

Supermarine Seafire Mk Ib (MB366)
801 Squadron, FAA
HMS *Furious*, Western Mediterranean, November 1942.

Supermarine Seafire Mk Ib (MB366)
801 Squadron, FAA
HMS *Furious*, Western Mediterranean, November 1942.

Supermarine Seafire Mk IIc (MB193)
880 Squadron, FAA
HMS *Argus*, Mediterranean Fleet, February 1943.

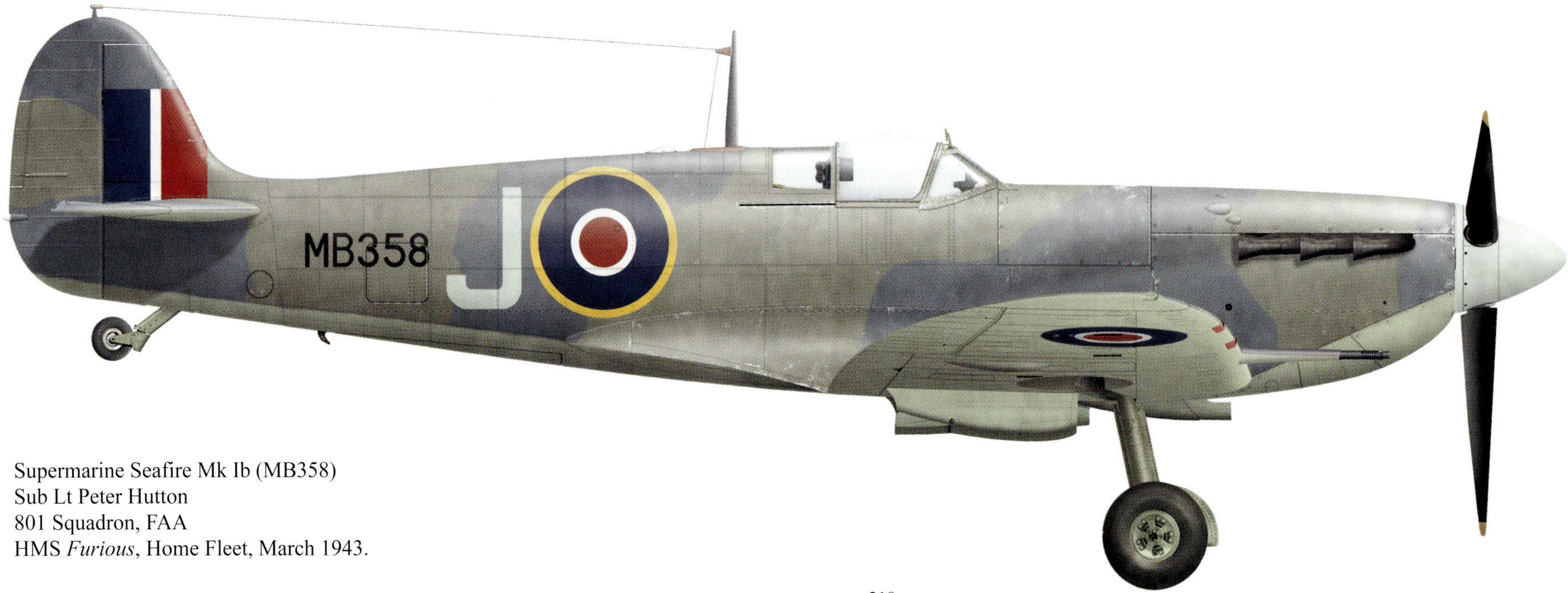

Supermarine Seafire Mk Ib (MB358)
Sub Lt Peter Hutton
801 Squadron, FAA
HMS *Furious*, Home Fleet, March 1943.

Supermarine Seafire Mk Ib (NX957)
761 Squadron, FAA
Henstridge, UK, March 1943.

Supermarine Seafire Mk Ib (MB345)
885 Squadron, FAA
HMS *Formidable*, Mediterranean Fleet, May 1943.

Supermarine Seafire Mk IIc (LR642)
Sub Lt R. H. Webber
807 Squadron, FAA
HMS *Battler*, Mediterranean Fleet, August 1943.

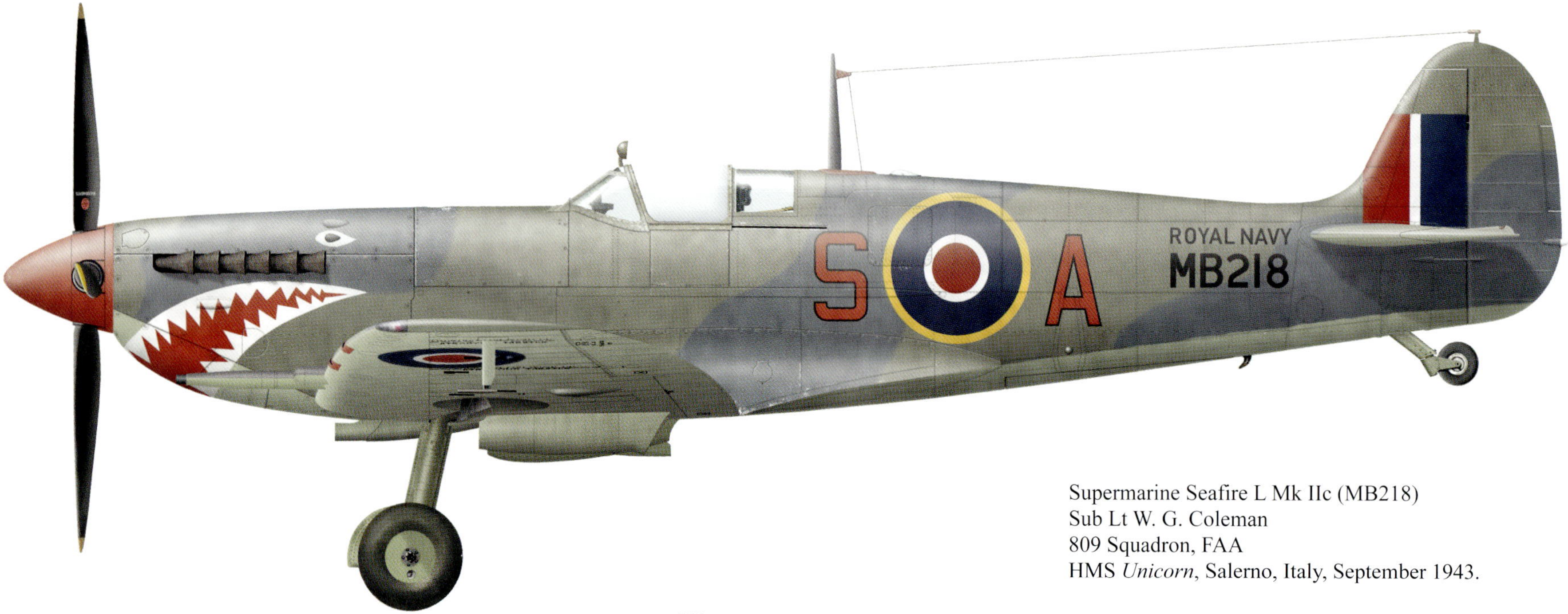

Supermarine Seafire L Mk IIc (MB218)
Sub Lt W. G. Coleman
809 Squadron, FAA
HMS *Unicorn*, Salerno, Italy, September 1943.

Supermarine Seafire Mk IIc (LR642)
Sub Lt R. H. Webber
807 Squadron, FAA
HMS *Battler*, Mediterranean Fleet, August 1943.

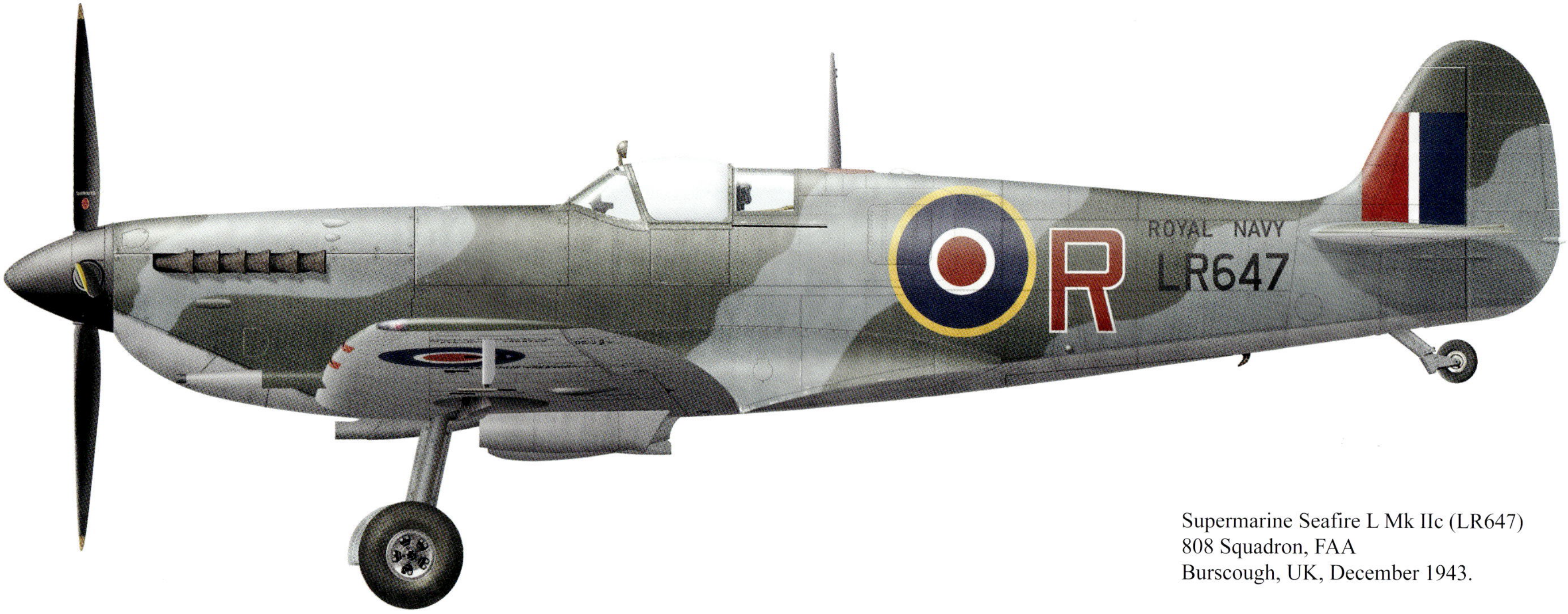

Supermarine Seafire L Mk IIc (LR647)
808 Squadron, FAA
Burscough, UK, December 1943.

Supermarine Seafire Mk Ib (MB349)
Lt Cdr Duncan Hamilton
Port Reitz Royal Navy Air Station
Mombassa, Kenya, February 1944.

Supermarine Seafire L Mk III (FN547)
Sub Lt R. C. S. Chamen
885 Squadron, FAA
Lee-on-Solent, UK, June 1944.

Supermarine Seafire L Mk III (NF541)
Lt Hugh 'Sam' Lang
886 Squadron, FAA
Lee-on-Solent, UK, June 1944.

Supermarine Seafire L Mk III (NN344)
899 Squadron, FAA
HMS *Khedive*, Agean Sea, summer 1944.

Supermarine Seafire L Mk III (NN460)
894 Squadron, FAA
HMS *Indefatigable*, East Indies Fleet, January 1945.

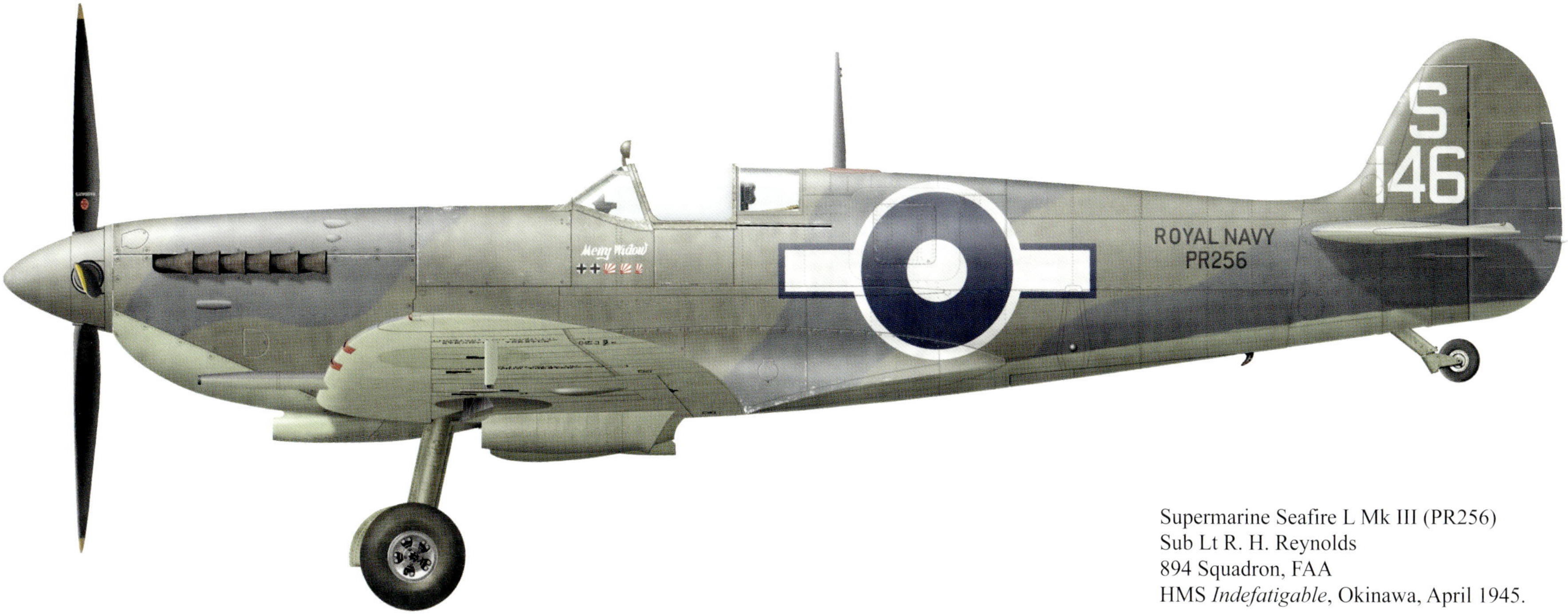

Supermarine Seafire L Mk III (PR256)
Sub Lt R. H. Reynolds
894 Squadron, FAA
HMS *Indefatigable*, Okinawa, April 1945.

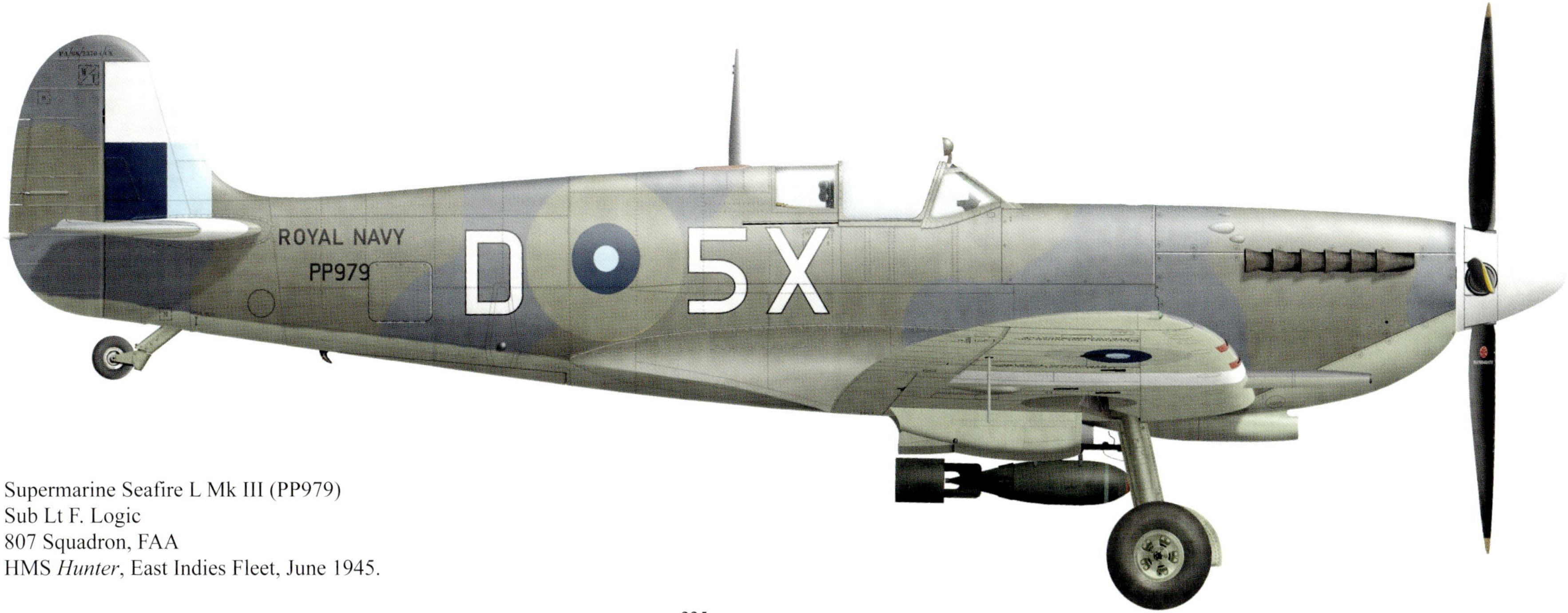

Supermarine Seafire L Mk III (PP979)
Sub Lt F. Logic
807 Squadron, FAA
HMS *Hunter*, East Indies Fleet, June 1945.

Supermarine Seafire L Mk III (NN621)
Lt Cdr Mike Crosley
880 Squadron, FAA
HMS *Implacable*, Sea of Japan, August 1945.

Supermarine Seafire L Mk III (NN300)
Lt Cdr George Baldwin
307 Squadron, FAA
HMS *Hunter*, Singapore, September 1945.

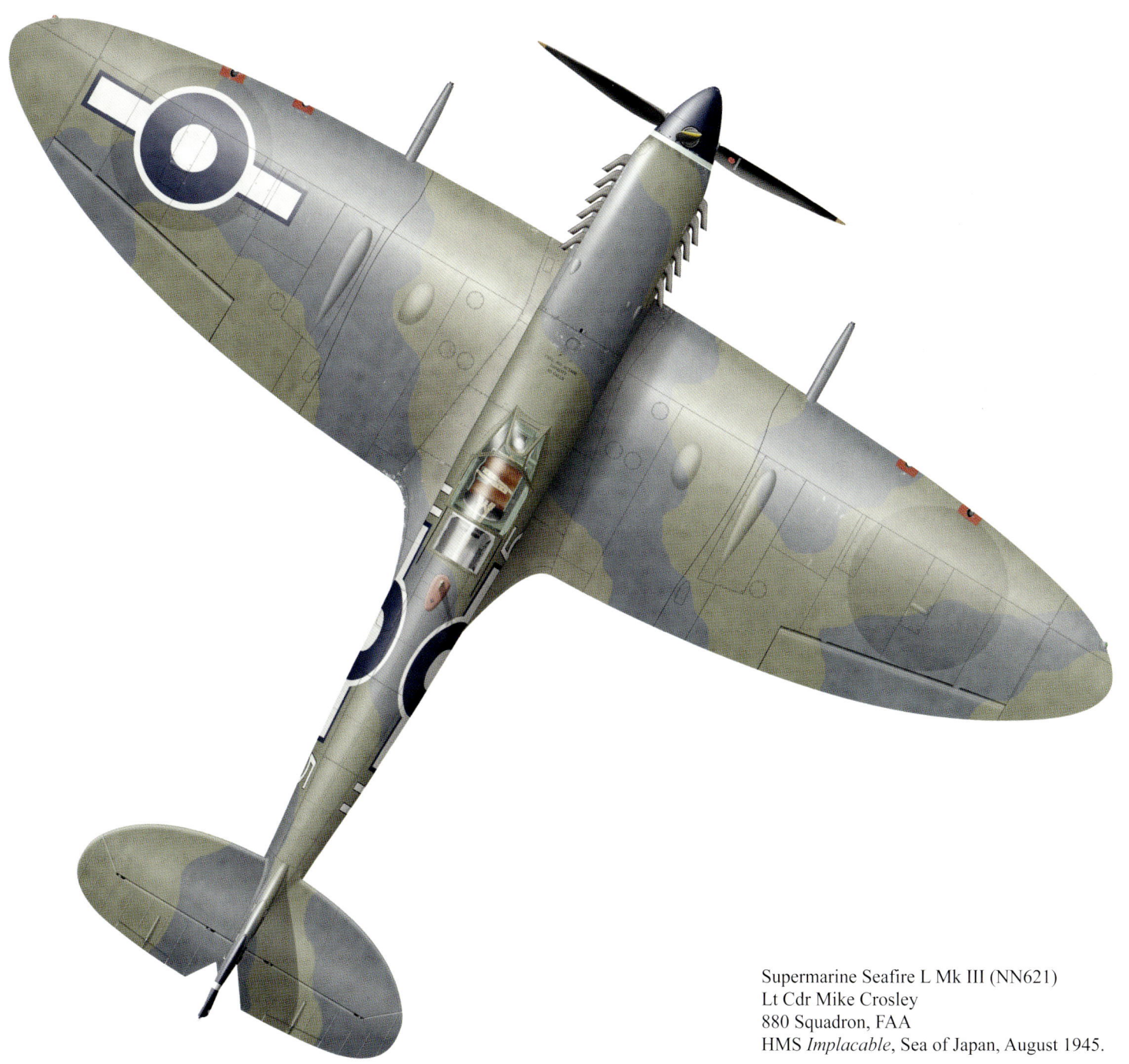

Supermarine Seafire L Mk III (NN621)
Lt Cdr Mike Crosley
880 Squadron, FAA
HMS *Implacable*, Sea of Japan, August 1945.

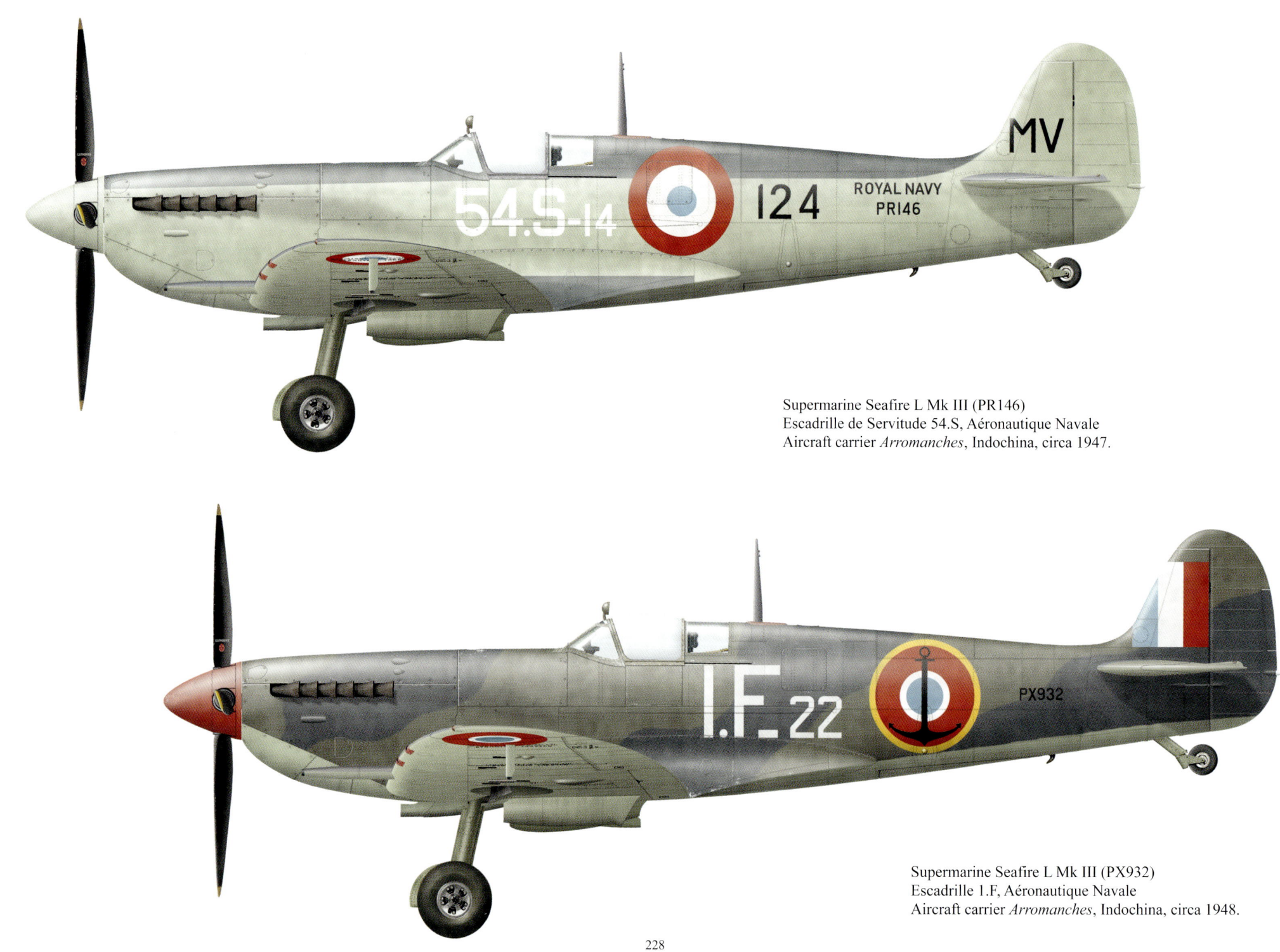

Supermarine Seafire L Mk III (PR146)
Escadrille de Servitude 54.S, Aéronautique Navale
Aircraft carrier *Arromanches*, Indochina, circa 1947.

Supermarine Seafire L Mk III (PX932)
Escadrille 1.F, Aéronautique Navale
Aircraft carrier *Arromanches*, Indochina, circa 1948.

RONNY BAR PROFILES